KU-050-717

FREE TO BE FOOLISH

FREE TO BE FOOLISH

POLITICS AND HEALTH PROMOTION IN THE UNITED STATES AND GREAT BRITAIN

Howard M. Leichter

PRINCETON UNIVERSITY PRESS PRINCETON, NEW JERSEY

Published by Princeton University Press, 41 William Street,
Princeton, New Jersey 08540
In the United Kingdom: Princeton University Press, Oxford

Library of Congress Cataloging-in-Publication Data

Leichter, Howard M.
Free to be foolish: politics and health promotion in the United States
and Great Britain / Howard M. Leichter.
p. cm.
Includes bibliographical references and index.
ISBN 0-691-07867-X (cloth : acid-free)
1. Health promotion—Political aspects—United States. 2. Health
promotion—Political aspects—Great Britain. 3. Health promotion
—Government policy—United States. 4. Health promotion—
Government policy—Great Britain. I. Title.
RA427.8.L45 1991
613'.0941—dc20 90-38310

This book has been composed in Linotron Sabon

Princeton University Press books are printed
on acid-free paper, and meet the guidelines
for permanence and durability of the Committee
on Production Guidelines for Book Longevity
of the Council on Library Resources

Printed in the United States of America by
Princeton University Press, Princeton, New Jersey

10 9 8 7 6 5 4 3 2 1

To Arlene Horodas Schwartz and Herman Schwartz

FOR EVERYTHING OVER ALL THESE YEARS

Contents

Figures

Tables

Preface

There is now widespread agreement among both the general population and health professionals that a good deal of disease is self-inflicted, the product of our own imprudent behavior. The premise that individuals contribute significantly to their own ill health or premature death appears unassailable in view of the mounting evidence relating various personal habits and life-style choices, such as poor nutrition, smoking, alcohol and drug abuse, failure to wear seat belts, and unsafe sexual practices, to major causes of morbidity and mortality.

While it is generally accepted that each of us is, to a certain extent, "dangerous to our own health," there is far less agreement on what can or should be done about making people less foolish. In particular, there is the question of how far government should go in refashioning life-styles to minimize the physical and mental harm we inflict upon ourselves and others in society through risky personal choices. Where does personal choice end and collective responsibility begin? How do we reconcile two of our most prized social values, personal freedom and good health? In this book I will be addressing these and related questions. I have chosen to examine them in a comparative study of British and American health promotion policies for two reasons. First, such a comparison can shed light on questions of enduring interest to political scientists, such as the relative impact of different national cultural and structural characteristics on the policy-making process. Second, public policy-making in general, and health promotion policy in particular, are inherently comparative enterprises. Many nations, and especially these two, often look to one another for policy inspiration.

There is something uniquely challenging about writing a book while on the faculty of a small liberal arts college. Library and other research support facilities are woefully inadequate; there are few, if any, colleagues on the faculty with a professional background in your particular field who can provide an audience for your misguided ideas; most of your time must be devoted to undergraduate teaching in areas outside of your research field. Under these conditions, the researcher/writer must rely to an unusual degree on the goodwill, patience, and generosity of others. I was especially exploitative and fortunate in this regard.

Since McMinnville, Oregon, is not the ideal location for comparative health policy research, I am indebted to the Northwest Area Foundation, the National Endowment for the Humanities ("Travel to Collections" grant program), and Linfield College for providing the time and financial support that allowed me to go where such research could be conducted. When I got "there," namely London, I was especially lucky to have the assistance of Bernard Weatherill, Speaker of the House of Commons, Toby Jessel, M.P., David Simpson of Action on Smoking and Health (ASH), and Michael Reddin of the London School of Economics and Political Science. I am most grateful to each of these gentlemen. In addition, the travel grant from NEH allowed me the opportunity to do research at the New York Academy of Medicine, which has one of the finest collections on the history of medicine in the country. The library staff there was most gracious.

However, it is closer to home that my debts are most numerous. No librarians have ever served more professionally and amiably than Lynn Chmelir, Michael Engle, and Frances Rasmussen of Northup Library, Linfield College. These extraordinary people stretched the meager resources of their library and, in one of the most creative and extensive uses of the interlibrary loan system ever undertaken, turned Northup into an ersatz research library. They were marvelous.

A number of colleagues at other institutions were kind enough to provide a long-distance forum for my ideas. Those who read all or parts of the manuscript and helped immeasurably in clarifying my thoughts include Michael Garland of the Oregon Health Sciences University; Donald Lutz, Malcolm Goggin, and Alan Stone of the University of Houston, and Deborah Stone of Brandeis University. Rudolf Klein of the University of Bath was especially thorough in his criticisms and helpful in his suggestions. I am most grateful, however, to Ted Marmor of Yale University, who was more generous with his time, more solicitous in his concern, and more perceptive and persistent in his criticisms than I had any right to hope for or expect. Among my colleagues at Linfield, I would like to single out Doug Cruikshank, who not only helped me master the mysteries of microcomputers and mathematics but remained remarkably calm throughout the ordeal. Finally, I would like to thank Gail Ullman of Princeton University Press for her gentle perseverance and support. Not even the collective wisdom and talent of all these extraordinary people could protect me from myself. Hence, I absolve all of them of any errors that remain.

I would like to thank the Hastings Center for permission to reproduce here material presented in my article, "Voluntary Health Risks and Public Policy: The British Experience," *The Hastings Center Report* 11 (October 1981): 32–39; and Duke University Press for permission to include material from my article, "Saving Lives and Protecting Liberty: A Comparative Study of the Seat-Belt Debate," *Journal of Health Politics, Policy and Law* 11 (Summer 1986): 323–44, copyright 1986 by Duke University Press.

As always the love, support, and sacrifices of my wife Elisabeth and my daughters Laurel and Alexandra made this effort both possible and worthwhile. I dedicate this book to Arlene and Herman Schwartz. I only hope that I can be as generous and loving to my children and grandchildren as they have been to theirs.

McMinnville, Oregon
April, 1990

Abbreviations

AMA	American Medical Association
BAC	blood alcohol concentration
BATF	Bureau of Alcohol, Tobacco, and Firearms
BMA	British Medical Association
CDC	Centers for Disease Control
EEC	European Economic Community
FTC	Federal Trade Commission
IOM/NAS	Institute of Medicine/National Academy of Sciences
JAMA	*Journal of the American Medical Association*
MADD	Mothers against Drunk Driving
NAACP	National Association for the Advancement of Colored People
NHSB	National Highway Safety Bureau
NHTSA	National Highway Traffic Safety Administration

FREE TO BE FOOLISH

One

Foolishness and Politics

IN 1808 a controversial but highly respected Scottish physician and professor of medicine at Edinburgh University, Dr. James Gregory, became involved in a debate over a parliamentary bill "To Prevent the Spreading of the Infection of the Small Pox." The bill would have prohibited vaccination within three miles of any city or town and required compulsory removal, isolation, and reporting of smallpox cases. The proposal met with considerable opposition, especially because of its compulsory provisions. Dr. Gregory, one of the bill's most vocal opponents, argued that "England is a free country and the freedom which every freeborn Englishman chiefly values is the freedom of doing what is foolish and wrong and going to the devil in his own way."[1] The bill was soundly defeated.

One hundred seventy-five years later, on the other side of the Atlantic, a nationally syndicated columnist, James Kilpatrick, argued against another compulsory, putatively health-promoting public policy, a rule proposed by the National Highway Traffic Safety Administration to require the installation of air bags in all new automobiles sold in the United States. Kilpatrick insisted that the decision on air bags should be a function of consumer demand, not government regulation. According to the conservative columnist, "a free people must be free to be foolish."[2]

Separated by nearly two centuries, thousands of miles, and vast technological changes, Dr. Gregory and Mr. Kilpatrick provided virtually identical answers to questions that historically have bedeviled pundits, policymakers, and the general public. Where does one draw the line between the power of the state to protect public health and the right of the people to make life-style choices? How much control should each of us have over decisions that may put ourselves, and perhaps others, at risk? Over the years, only the specific content of this debate has changed—compulsory vaccination against smallpox, or compulsory fluoridation; mandatory notification of tuberculosis cases, or of AIDS cases; safe milk, or safe sex; child nutrition laws or minimum drinking-age laws. Whatever its specific form, the essential issue remains the same: What is the appropriate role of the state in promoting responsible personal behavior?

[1] Quoted in Bernard J. Stern, *Should We Be Vaccinated?* (New York: Harper and Brothers, 1927), 48.

[2] Portland *Oregonian*, 23 December 1983.

Periodically this issue has taken a prominent position on the political stage, as was the case in the decades just prior to and succeeding the turn of the last century and is the case today.

This book is about the current debate, and seemingly constant tension, over the role of government in two postindustrial democratic societies, the United States and Great Britain, in helping people secure both good health and personal freedom. I will argue that in recent years there has been a fundamental shift in the prevailing wisdom among opinion leaders and policymakers in both countries about the way people can best achieve healthy lives. The shift has resulted in increased emphasis on health promotion, especially through government-sponsored life-style modification, and decreased attention to securing access to health care. Over the past decade the political agendas of most postindustrial, democratic societies have been crowded with proposals to promote more responsible drinking, diet, driving, and sexual and substance-use behavior. These issues have transformed health politics from a largely consensual to a largely conflictual enterprise. This transformation results from the fact that health promotion policies differ significantly, in ways to be described below, from other areas of public policy and, indeed, other types of health policy.

Although health promotion issues play a prominent role in the personal lives and collective concerns of people in the United States and Great Britain, each nation has responded somewhat differently to the political and medical challenges posed by such public health problems as AIDS, tobacco and alcohol use, and motor-vehicle driving. A primary concern of this book, then, is to account for the similarities and differences in health promotion policy in the United States and Great Britain.[3] I will suggest that despite common public health challenges, variations in cultural values and political structure have produced differences in the timing, tempo, and nature of the health policy response in each country.

A New Perspective on Health

It is difficult to conceive of anything more important or basic to human happiness, dignity, or freedom than good health. Over two thousand years ago the Greek physician Herophilos said, "When health is absent, wisdom cannot reveal itself, art cannot become manifest, strength cannot

[3] In this book I use the World Health Organization definition of health promotion as "the process of enabling people to increase control over, and to improve, their health." *Health Promotion: A Discussion Document on the Concept and Principles* (Geneva: W.H.O., 1984).

fight, wealth becomes useless and intelligence cannot be applied."[4] Centuries later Ralph Waldo Emerson wrote, "Give me health and a day and I will make the pomp of emperors ridiculous."[5]

The importance that people in the United States attach to sustaining good health is reflected in the billions of dollars we spend each year directly on health care—an estimated $590 billion in 1989, or $2,306 per person—and indirectly on things like vitamins, health foods, exercise, guides to health, and other paraphernalia, all in what one British observer has called America's "national preoccupation with health."[6] A Gallup poll reported that among the goals most sought by Americans, the two most frequently cited were a good family life (82 percent) and good physical health (81 percent), while another study found that 42 percent of the respondents "think about their health more often than just about anything else, including love, work, and money."[7]

Although Americans are by no means alone in the pursuit of good health, impressionistic and empirical evidence suggests that we may be more obsessive than most. For example, one study found that 97 percent of the people surveyed in the greater Cleveland area reported doing "something" (following a healthy diet, getting enough sleep, exercise, or relaxation, and so on) to maintain or improve their health. British researchers asked an identical question of a national sample and found that only 60 percent of their respondents reported similarly health-protective behavior.[8] There are obvious sampling and cultural reasons why these two studies may not be entirely comparable, but the findings support anecdotal and descriptive data suggesting that the British are not as ardent in the pursuit of healthy life-styles as Americans. Nevertheless, the proliferation of health-food shops, fitness centers, and recreational magazines, along with the Thatcher Government's emphasis on personal responsibility, suggests that the gap may well close in the future.

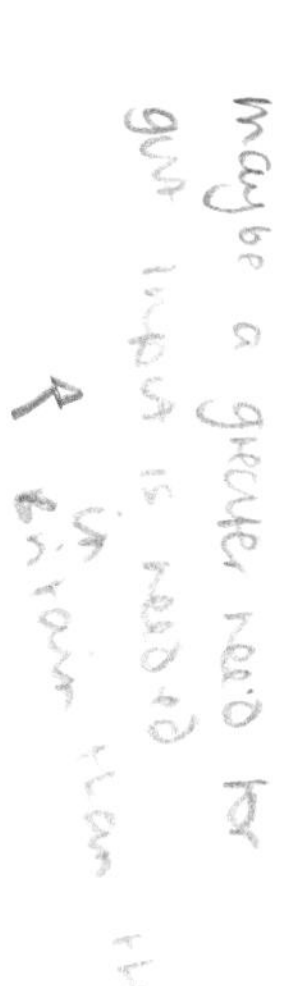

In addition to the importance each of us places on good health, there is a special place assigned by democracies to the physical well-being of their citizens. Marvin Bressler calls this commitment one of our "master-val-

[4] Quoted in Henry E. Sigerist, *Medicine and Human Welfare* (New Haven: Yale University Press, 1941), 57.

[5] Quoted in Harry H. Moore, *Public Health in the United States* (New York: Harper and Brothers, 1923), 2.

[6] Christopher Potter, "Showdown for Health Vigilantes," *Health and Social Services Journal* (22 September 1983): 1140.

[7] Carin Rubenstein, "Wellness is All," *Psychology Today* (16 October 1982): 28–37.

[8] See Daniel M. Harris and Sharon Guten, "Health-Protective Behavior: An Exploratory Study," *Journal of Health and Social Behavior*, no. 1, 20 (1979): 17–29; and Mildred Blaxter, Nigel Fenner, and Margaret Whichelaw, " 'Healthy' Behaviour," in, *The Health and Lifestyle Survey*, ed. Mildred Blaxter et al. (London: Health Promotion Research Trust, 1987), 121.

ues": "democratic societies are, by definition, committed to a series of ethical standards emphasizing the value of human life and well-being."[9] It is not surprising that historic political leaders have conspicuously displayed their commitment to the public's health. Benjamin Disraeli once said that, "Public health is the foundation on which reposes the happiness of the people and the power of a country. The case of the public health is the first duty of a statesman."[10] Clearly, then, good health plays a prominent role in both the individual and collective value systems of democracies such as the United States and Great Britain. All this is not to suggest that the whole point of living is to stay alive regardless of the sacrifice in personal pleasure or freedom. Skiers, mountain climbers, and smokers, to name a few, all make a trade-off between the statistical possibility of harm and the apparent certainty of pleasure derived from their activities. Having recognized this, I would return to my original point about the critical role of good health as a social ideal. To my knowledge there never has been any serious political debate over the desirability of being and staying healthy.

There has been considerable debate, however, over the question of how people and nations can best achieve what they so highly prize, and at what cost. There is, to begin with, the basic issue of the appropriate role of government in the area of health. Although it has long been accepted by people of all political persuasions that the state has some role in protecting the health of its people, there is less agreement over the precise nature and extent of that role. In terms of public policy, there are, in effect, four ways by which the state can help further the health of its people. It can support biomedical research; improve, guarantee, or subsidize access to health care; regulate environmental and product hazards; and encourage people, through education or regulation, to adopt more healthy life-styles.[11]

For much of the last century or so, in both Britain and the United States, the prevailing wisdom has been that people may best maximize their prospects for good health through access to adequate health care. As a result, the emphasis in national debates on health policy has been on health care delivery. In this context, one major issue has been whether or not the state should help people secure access through a national health insurance system. Ultimately the British adopted a limited health insur-

[9] Marvin Bressler, foreword to "Meeting Health Needs by Social Action," *Annals of the American Academy of Political and Social Science* 337 (1961): ix.

[10] Quoted in Moore, *Public Health*, 2.

[11] These four approaches parallel the "Health Field Concept" developed by Marc LaLonde, *A New Perspective on the Health of Canadians* (Ottawa: Department of National Health and Welfare, 1974); see also James W. Vaupel, "Early Death: An American Tragedy," *Law and Contemporary Problems* 40, no. 4 (Autumn 1976), 74–121.

ance system in 1911 (it began operating in 1913), and then a comprehensive National Health Service (NHS) in 1948. In the United States it has been decided that only certain groups that are vulnerable (for example, the needy and the aged) or special (for example, veterans) should receive direct assistance from the state in getting health care, although many more people receive indirect assistance through tax deductions on private health insurance. Although there has been increased public funding of biomedical research in both countries over the years, in neither has this been an area of major state emphasis. In the United States, for example, government spending on biomedical research since the end of World War II generally has been around 2 percent of total national health expenditures.

In the past decade there has been an important shift in the debate over how people can maximize their chances of staying healthy. Since the mid-1970s, in most industrialized countries, greater emphasis has been placed on reducing environmental health hazards and encouraging more prudent life-styles. One of the earliest statements of policy support for this shift came in a 1974 Canadian government working paper, "A New Perspective on the Health of Canadians," by then minister of health and welfare, Marc Lalonde. The Lalonde report identified the major causes of morbidity and premature mortality in Canada (motor-vehicle accidents, ischemic heart disease, respiratory diseases, lung cancer, and suicide) and attributed them to self-imposed risks, that is, life-style, and environmental factors, including various forms of pollution. Thus, the Lalonde report placed life-style and the environment on a par with the more traditional concerns of health policy such as human biology and health care delivery.

The "new perspective" on health, as it came to be known, heralded a new era in which health care professionals, public policymakers, academics, and the general public have come to believe that greater progress toward our becoming a healthy people can be made through reducing both environmental hazards and self-indulgent, health-endangering personal behavior than in expanding access to health care.

Despite considerable scholarly concern with the validity and implications of the "new perspective" (see Chapter 3), few would challenge the assertion that an important shift has occurred in both the United States and Great Britain, in debates over health policy. Because of its national health system, Britain has continued to place considerable policy emphasis on health care delivery, but even there both Labor and Conservative governments since the 1970s have argued that prevention, especially through life-style modification, is the most cost-effective and medically efficacious way of improving the nation's health. In the United States, the shift at the policy-making level has been more pronounced, one crude measure of which is the virtual disappearance of the debate over national health insurance at the national level. For example, in 1970 and 1971 the *Congressional Record* lists a total of fifty references to national health insurance; between 1981 and 1988 there are three. Although there has been a recent revival of interest at the state level in guaranteeing access to health care for those without health insurance, the preponderance of public and private concern remains in promoting healthy behavior rather than expanding access. In both countries, policymakers and administrators have shifted their attention to such measures as mandatory seat-belt and minimum alcohol drinking-age laws, restrictions on the sale, promotion, and advertisement of tobacco and alcohol products, and distribution of free condoms and clean hypodermic needles, as well as various environmental health concerns. Health politics since the 1970s has been, to a considerable extent, the politics of health promotion and disease prevention.

Although environmental and life-style strategies are at the core of the current policy emphasis on health promotion, the two differ in their politics and political and social implications. Environmental policies tend to focus on collective, especially corporate, behavior such as industrial waste disposal, water, air, and noise pollution, and workplace hazards, while life-style policies seek to influence individual choice.[12] The distinction between individual and collective behavior may not always hold, but it is an analytically useful one. This book will focus on those policies that seek to change individual, rather than corporate or collective, behavior; although in accomplishing the former (for example, reducing alcohol and tobacco consumption) they sometimes must do the latter (for example, requiring manufacturers to use health warning labels).

By excluding environmental and workplace health issues from this discussion, I am, of course, slighting a vital dimension of the overall health promotion agenda. The decision to do so was based on both tactical and analytical considerations. It seemed impractical to tackle, in a single volume, the full range of health promotion issues, especially since, as I argue below, life-style and health politics and policies differ significantly from those dealing with the environment or the workplace. In fact, and this is the more important reason for devoting my attention to life-style and health, these issues, more so than even environmental degradation or hazardous work conditions, have become politically emblematic of our time. Politically in the last decade, in both the United States and Britain, national administrations have sought to reduce the role of government in our lives and encourage greater self-reliance. In the area of health, this has meant shifting much of the responsibility for promoting health and

[12] See Michael R. Pollard and John T. Brennan, "Disease Prevention and Health Promotion Initiatives: Some Legal Considerations," *Health Education Monographs* 6 (1978): 211.

the culpability for endangering it from the collective, both private and public, to the individual arena. This is not to suggest that environmental or occupational health problems have been ignored or that they are unimportant, but simply that the philosophical and political emphasis in the Thatcher and Reagan-Bush governments has been on individual rather than corporate actors.

Conterminous with the political shift from a collectivist to an individualist political orientation has been a change in the health consciousness of people in Britain and, especially, the United States. Beginning in the late 1970s in both countries, particularly among the younger, more affluent, and politically influential, the pursuit of good health through life-style change became a major personal preoccupation. In the United States, for example, the proportion of people indicating that they do something regularly to keep physically fit more than doubled between 1961 and 1984, from 24 to 59 percent of the population. In addition, the proportion of the population reporting that most prototypically health-pursuing behavior, namely jogging, increased over the same period from 6 to 18 percent.[13]

In sum, both prevailing political sentiment and social habits reflected and inspired emphasis on the individual and the importance of personal responsibility. In this context life-style and health issues virtually became a symbol for our times.

Health Promotion, Politics and Personal Freedom

It is a central thesis of this book that public policies seeking to alter individual life-styles differ from policies on health-related environmental and workplace issues, and indeed other areas of health policy, in the nature of the politics and the range of policy alternatives they produce. Let me be more specific about this point.

To begin, given what I have said already about the importance attached to good health, one might expect public efforts to promote health to enjoy enthusiastic political support. As Theodore Litman has noted, health politics "is usually conducted in a favorable political climate. The notion of health is a popular one."[14] However, as anyone who has followed the public debates over mandatory seat-belt laws, drug and AIDS testing, or

[13] See the Gallup Poll, *Public Opinion 1935–1971*, vol. 3 (New York: Random House, 1972), 1743, and the Gallup Poll, *Public Opinion 1984* (Wilmington, Del.: Scholarly Resources, 1985), 112.

[14] Theodore J. Litman, "Governmental Health: The Political Aspects of Health Care—A Sociological Overview," in *Health Politics and Policy*, ed. Theodore J. Litman and Leonard S. Robins (New York: John Wiley and Sons, 1984), 31.

limiting smoking in public places knows, these issues have been debated in a political climate that hardly can be called "favorable." Health promotion policies, especially those involving modification of personal behavior, are unlike health-related policies involving biomedical research, hospital construction, or health care for the aged. Health promotion and disease prevention constitute a special species in the health policy genus, for a number of reasons.

The first stems from the fact that many proposed health promotion practices involve our most basic, intimate, and routine personal behavior patterns, including the quantity and nature of the products we consume (such as food, alcohol, drugs, and tobacco), and with whom and how we engage in sex. The highly intimate nature of many of these proposals, and hence one reason why they produce such extraordinary politics, was underscored by J. Michael McGinnis, a deputy assistant secretary for health in the Reagan administration: "Efforts to prevent disease and promote health get us into areas that affect individual citizens in much more personal ways than do decisions about tax policies or foreign trade or national defense. Disease prevention and health promotion involve decisions affecting not only a person's pride and pocketbook but also his or her body, soul, and psyche."[15]

The intimate and often intrusive nature of so many proposed health promotion policies historically has made this an area of extraordinary policy passion. Certainly one can appreciate those in the nineteenth century who opposed a compulsory preventive health procedure described in the following manner: "Vaccination is the cutting, with a sharp instrument, of holes in your dear little babe's arm, and putting into the holes some filthy matter from a cow—which matter has generally in addition passed through the arm of another child. So that your dear little babe, just after God has given it you, is made to be ill with a mixture of the corruption of both man and beast, which is forcibly inserted into its body."[16]

Although most modern preventive health measures such as mandatory seat-belt laws do not involve the same degree of physical violation characteristic of vaccination or nineteenth-century examinations for venereal disease—drug and AIDS testing comes close—for many the principle is the same: What people do with their own bodies should be a personal, not public, decision. Arguing against a proposed mandatory seat-belt law, a New York State legislator beseeched his colleagues to "leave me alone; let me make my own decisions."[17]

[15] J. Michael McGinnis, "The Limits of Prevention," *Public Health Reports* 100 (May-June 1985): 257.

[16] *Vaccination Tracts*, no. 6 (London: William Young, 1879).

[17] New York State Senate proceedings (25 June 1984), 5236–37.

Second, life-style-related health promotion policies generate extraordinary passions, and produce unique policy conditions, in part because of their explicit assumption about the etiology of disease. Many of today's so-called "life-style diseases" are conceptualized as being self-inflicted, the result of foolish personal behavior, rather than of random and uncontrollable biological forces. There is the tendency, most notably in the case of AIDS, to think in terms of guilt or innocence when talking about persons with a "life-style disease." Many gays, for example, have criticized the media's portrayal of nonhomosexuals with AIDS as "innocent victims," as if homosexuals were "guilty," and therefore justifiably ill.[18] This would not be the first time, of course, that people have been presumed culpable for their own ill health (see Chapter 2).

A third, related factor distinguishing health promotion from other policies is the tendency to view the health problems involved as not only self-inflicted but the result of a "weak character," which our society scorns or, worse still, a sign of immorality. David Mechanic has observed that the motivation of some who wish to change health-diminishing behavior is often associated with moral revulsion rather than health concerns: "Those people who wish to reform the unhealthful habits of others give as their stated motive for changing certain habits the damage caused by the habits upon the individual's health. But their real reason—conscious or otherwise—might be simple moral revulsion to the habits themselves."[19] Contemporary reactions to AIDS, which was unknown when Mechanic made this observation, and even smoking-related illnesses corroborate this point. The attribution of illness to immorality or character weakness is, once again, reminiscent of an earlier time. Nineteenth-century reactions to alcoholism, venereal disease, and even cancer were often couched in terms of divine retribution or personal failing. One hundred years ago a British physician, testifying before a parliamentary committee on the Contagious (i.e., venereal) Diseases Act, said that syphilis and gonorrhea were "the penalty which it had pleased God to inflict upon the commission of vice."[20] This comment might just as easily have been made yesterday about AIDS.

The point I wish to emphasize is that when health policy takes on the quality of a moral or social crusade, as it often does, the nature of the policy debate, and its outcome, differ from that of other political debates. How society characterizes a particular health problem influences the re-

[18] See Dennis Altman, *AIDS in the Mind of America* (Garden City, New York: Anchor Press, 1986).

[19] Quoted in Daniel I. Wikler, "Persuasion and Coercion for Health: Ethical Issues in Government Efforts to Change Life-Styles," *Milbank Memorial Fund Quarterly* 56, no. 3 (1978): 14.

[20] Select Committee on Contagious Diseases Act (2 May 1882), 211.

sponse to that problem. Alcoholism, for example, has been defined variously as a disease, a moral weakness, or the result of socioeconomic oppression. Society will support different policy responses to "sinners" than to the "sick" or "oppressed." Furthermore, accepted definitions of health problems change over time. Once considered immoral behavior, alcoholism is now widely viewed as a disease, although this idea too has been challenged recently. Similarly, smoking is moving from the realm of an accepted, indeed glamorous, activity, to one of strong social disfavor. This shift is, I think, partly explained by the fact that smoking, much as alcohol consumption in the latter part of the nineteenth century, has become increasingly identified as a lower-class activity. It has always been easier to recognize personal weaknesses in socially marginal groups than in the "better" classes. There is a great deal of moral posturing in the health promotion debate.

It is important to note that such posturing is often deliberate and politically functional. Because many health promotion policies seek either to restrict individual behavior like smoking or drinking or control a social setting by, for example, creating safer roads or cleaner air, they often stimulate fierce opposition by preexisting interest groups such as automobile, tobacco, and alcohol manufacturers and producers. One way of overcoming such opposition is to change the idiom of the debate from economic regulation to protection of public health through prevention of immoral or antisocial behavior.[21]

Fourth, health promotion and disease prevention policies also generate extraordinary passions and politics because they intrinsically conflict with our individual freedom. This conflict is a timeless one. Consider, for example, an early variation of the "better dead than Red" theme. In an article opposing compulsory vaccination, Millicent Fawcett, a leading nineteenth-century feminist wrote, "I am one of those who sympathise with the bishop who would rather see England free than sober; and how much more, therefore, do I feel that I would rather see England pock-marked than without the personal independence which is the basis of everything worth having in our national character."[22] A more recent expression of the conviction that it is better to suffer the consequences of (allegedly) reckless behavior than to submit to coercive state regulation was made by a Washington State legislator who, speaking against a mandatory seat-belt bill in his state, said, "There is something more important than life itself, and that's freedom."[23]

It is instructive in this regard that, after years of denying any link be-

[21] I am indebted to Theodore Marmor for this point.

[22] Millicent Fawcett, "The Vaccination Act of 1898," *Contemporary Review* 75 (1899): 334.

[23] Senator Kent Pullen, Washington State Senate (17 April 1985).

tween smoking and ill health, and then challenging the credibility of the scientific evidence supporting this link, the American tobacco industry has finally switched its defense of smoking to the issue of personal freedom. Beginning in 1988, the Tobacco Institute launched its "Enough is Enough" campaign aimed at convincing nonsmokers that smokers were the victims of censorship, government harassment, public control of private behavior, and discrimination. This new message, along with state record-breaking spending on a ballot measure, resulted in the defeat of an initiative on the Oregon Indoor Clean Air Act in November 1988.

In sum, health promotion policies often require a choice between good health and personal freedom. Health promotion thus becomes a quintessential political issue, involving intense conflict and requiring difficult policy choices. One of the tasks of this book is to help identify some of the criteria used by policymakers in Britain and the United States in making these choices.

A fifth reason for the highly charged nature of most health promotion debates is that individual life-style choices are typically portrayed as having consequences that extend beyond the individual concerned. These consequences may place others at physical risk, or cause them material deprivation, or limit their freedom. To the extent that such activities are "other-regarding," to use a phrase introduced by John Stuart Mill, or produce negative "externalities," a concept derived from economics, they arouse deep political concerns. Referring to a variety of health promotion policies including fluoridation, motorcycle helmet laws, and inoculation against communicable diseases, Donald Kennedy, a former commissioner of the Food and Drug Administration and current president of Stanford University has said, "These matters are difficult precisely because they raise basic questions of social justice. Such questions must be approached through the analysis of how particular actions affect the welfare of others in society—or, in other words, by looking at the distribution of risks and benefits that resulted from whatever action we seek to regulate."[24]

The point is that individual life-style choices become socialized, and often assessed not merely for the good or ill they bring to the individual concerned but also for their collective implications. This, too, is by no means a phenomenon unique to our own time. The basis for judging the social relevance of personal life-style and behavioral choices has varied over time and been defined by contemporary need. In times of national crisis, such as war, personal imprudence has been viewed as endangering national security. This occurred during World War I, when public officials feared an epidemic of venereal disease among the U.S. Expeditionary

[24] Donald Kennedy, "Health, Science, and Regulation: The Politics of Prevention," *Health Affairs* 2 (Fall 1983): 46.

Force. In 1917, Secretary of the Navy Josephus Daniels, in a widely distributed pamphlet entitled *Men Must Live Straight If They Would Shoot Straight*, wrote that "we are fighting for the safety of democracy. Victory is jeopardized by the preventable diseases which destroy the fighting strength of armies and navies." The secretary reminded fighting men that democracy carried with it certain obligations: "Those obligations require the individual to curb his passions and exercise self-restraint in order that the institution of the family, which is the fountain-head of the State, and from which springs all our noblest inspirations, shall remain pure and undefiled."[25] Irresponsible sexual behavior threatened not merely the health of the individual doughboy, but the institution of the family and democracy itself. Although the relationship between personal recklessness and social well-being has not always been as explicitly or dramatically portrayed, it is often an important element in health promotion debates.

Another, more recent, illustration of the socialization of personal imprudence is the current emphasis on the relationship between individual life-style choices and the national explosion in health care costs. John Knowles, former president of the Rockefeller Foundation, once noted that "the cost of sloth, gluttony, alcoholic intemperance, reckless driving, sexual frenzy and smoking have now become a national not an individual responsibility. One man's or woman's freedom in health is now another man's shackle in taxes and insurance premiums."[26] Knowles's view was echoed by Joseph Califano during his tenure as secretary of health, education, and welfare. Califano wrote in the foreword to the 1979 surgeon general's report *Healthy People* that "indulgence in 'private' excesses has results that are far from private. Public expenditures for health care that consume eleven cents of every federal tax dollar are only one of those results."[27] It is because virtually all life-style choices impose some costs on persons not directly engaged in the behavior that these issues pose such an extraordinary policy dilemma. Clearly the government cannot intervene in every instance in which personal behavior has the potential to cause discomfort, harm, or cost to others. Aside from the obvious impracticality of such intervention, it is important to recognize that regulating what may be offensive or injurious to some produces a reciprocal imposition of costs, or deprivation of freedom, for still others. A ban on

[25] Josephus Daniels, *Men Must Live Straight If They Would Shoot Straight* (Washington, D.C.: Navy Department, 1917), 16.

[26] Quoted in Robert Crawford, "Individual Responsibility and Health Politics in the 1970s," in *Health Care in America*, ed. Susan Reverby and David Rosner (Philadelphia: Temple University Press, 1979), 255. Emphasis added.

[27] United States Surgeon General, *Healthy People* (Washington, D.C.: Public Health Service, 1979), ix.

smoking in public places may lead to greater comfort for nonsmokers, but it also will produce discomfort and injury (loss of pleasure, increased anxiety) to smokers. Clearly the rights, benefits, and costs of all parties must be weighed in the lawmaking process.

Having argued that life-style politics are unique, I would hasten to add that it does not necessarily follow that each debate over changing personal behavior is in some sense an identical, predetermined set battle. Although policies dealing with AIDS and automobile safety each produce uncommon passions, as policy issues they engage different political structures and processes and produce different policy configurations. Nevertheless, there is a distinctive quality about health promotion politics that warrants separate analysis.[28]

Political Standards for Evaluating Health Promotion Policies

The factors that make health promotion policies so extraordinary produce a series of questions that arise each time there is a proposal to regulate personal behavior in the name of health promotion. When considered together, these questions provide a working evaluative framework by which the appropriateness of life-style modification policies are and should be judged. The questions, which will be revisited throughout this book, deal with issues of evidence, etiology, externalities, and alternatives.

The Nature of the Evidence

Among the most basic questions that must be addressed before the public and policymakers can and should endorse regulation of personal behavior in the name of health promotion is the degree of certainty linking a particular life-style practice to a health harm. Is the practice a necessary and sufficient *cause* of the harm, or is it simply associated with it? What is the nature of the evidence linking the life-style practice and the health harm? What standards of evidence should be required before a causal or associational relationship is accepted as valid? How unambiguous is the relationship? For example, does cigarette smoking cause lung cancer in the same way that a bowling ball causes pins to fall down? Or is cigarette smoking one of several factors associated with the increased risk of con-

[28] For an admonition about the dangers of treating all health politics as an undifferentiated mass, see Theodore R. Marmor and Amy Bridges, "Introduction to Special Issue: Politics, Medicine, and Health," *Journal of Health Politics, Policy and Law* 4 (Fall 1979): 354–59.

tracting lung cancer? Does the data linking smoking and ill health prove a causal relationship, or merely a statistical association? Is the evidence of association sufficiently compelling to warrant public policy that inhibits or penalizes the use of cigarettes?

The fact that citizens, or even policymakers, rarely can independently corroborate the purported relationship between a style of living and a physical harm makes the evidence itself an integral part of the politics of health promotion. Arguing that the early onset of coitus, or multiple sexual partners, increases the risk of cervical cancer is not something most people feel qualified to judge as they might, for example, whether a tax cut increases consumer spending. This problem is generically characterized by Duncan MacRae, Jr., as "the inequality of knowledge between scientists and other citizens."[29]

Another dimension of the evidentiary issue is the one linking a proposed policy to a desired behavioral change. What is the evidence, for example, that raising taxes, or banning advertising, or limiting public use of cigarettes will achieve the goal of reducing smoking? What is considered statistically or descriptively acceptable evidence before adoption of a policy that is intended to affect behavior?

The first set of questions, then, deals with the standards of evidence upon which health promotion policies should be based. Quite often the validity of the evidence used in health promotion policy debates is seen by both opponents and proponents as a threshold issue, facilitating or inhibiting subsequent policy efforts.

The Etiology of Harm

Closely related to the nature of the evidence linking behavior to harm is the issue of the origins or etiology of disease, injury, and illness. Is a particular harm the result of informed, voluntary choice, or factors beyond the individual's control? Is, for example, alcoholism a consequence of personal irresponsibility or genetic predisposition? Is AIDS a "no-fault" disease for homosexual males who engage in anal sex in the same way that, say, toxic-shock syndrome is for females who use tampons?

In terms of public health promotion policy the critical question that follows is whether or not policymakers should distinguish between self-inflicted and involuntary diseases. Should public policymakers show greater compassion for and tolerance of, and allocate greater resources

[29] Duncan MacRae, Jr., "Science and the Formation of Policy in a Democracy," *Minerva* 11 (1973): 229.

to, those who have not brought disease and injury upon themselves than those who knowingly court danger through their behavior? When this facet of an issue is debated, the language of guilt or innocence is frequently introduced.

Externalities

Externalities are costs or benefits borne by individuals not directly engaged in an activity. These costs may be tangible and material (for example, increased insurance rates or higher highway safety costs), or they may be intangible (for example, discomfort, ill health). Externalities pose a substantial dilemma for regulatory policy in general, and health-promotion policy in particular. The general problem has been well stated by Alan Stone: "Almost every act in a complex, crowded, industrial society involves externalities, but we would not expect government to institute rules for all of them. Consequently, an important task is to determine which externalities might be candidates for government intervention and which should clearly be ruled out."[30]

How does one determine which externalities require government intervention? One set of critical evaluative questions involves the extent and nature of the alleged harm. Thus, one might ask how many are negatively affected by a particular behavior pattern? Stone, for example, suggests the quantitative standard that "an externality whose probable impact is restricted to one or a small group does not warrant public action."[31] The difficulty here is, of course, that this does not differentiate among consequences. What if the impact harms only a few but does so seriously? Should we not measure the nature, as well as the extent, of the harm?

Similarly, what about the nature of the externalities? Do we want to distinguish between, for example, behavior that imposes material costs such as road repair, law enforcement, and health care costs associated with an automobile accident, and behavior resulting in physical harm? In addition, and this is related to the question of evidence and causality, should policy be based upon the distinction between actual externalities and potential externalities?[32] Some may wish to argue that regulation is only justified when actual harm occurs.

Questions of externalities permeate debates over health-promotion policy. Because, as Stone has noted, virtually everything we do creates

[30] Alan Stone, *Regulation and Its Alternatives* (Washington, D.C.: Congressional Quarterly Press, 1982), 91.

[31] Ibid., 98.

[32] Ibid., 100.

some cost (or benefit) for others, the question of where to draw the line is perhaps the one most frequently asked in the course of health policy debates.

Viable Alternatives

Health promotion regulatory policy often requires people to forsake or limit a particular activity or raises the costs or difficulty of engaging in that activity. Because of the freedom-limiting implications of so many health promotion measures, questions are often raised concerning both public and private policy alternatives to proposed regulations. Short of regulation, how can a desired, health-enhancing outcome such as increased use of seat belts, more prudent use of alcohol, or reduced use of cigarettes be achieved? Should policymakers follow a hierarchy of possible responses, ranging from public exhortation and education to disincentives like increased taxes before resorting to regulating behavior?

An additional set of related questions involves whether the desired outcome could not be better achieved through private rather than public means. Since so many of the externalities involved affect private insurance rates, would not this be a more effective venue for making people more prudent, without expanding the role of government? Or, to use another example, if many of the external costs associated with smoking, including decreased productivity, higher absenteeism, and larger health insurance costs, affect private business, would it not make more sense to allow market forces and business practices to create incentives for reducing smoking in the workplace, rather than having government mandate it? Certainly in the context of both American and British disillusionment with government-sponsored regulation, private and voluntary efforts at health promotion appear especially appealing.

There are a finite number of policy choices available in pluralist democracies for dealing with life-style and health issues. As I have suggested above, policymakers confront questions dealing with the nature of the evidence linking life-style and disease, the etiology of disease, the externalities incurred as a result of both imprudent personal behavior and policies seeking to alter that behavior, and available policy choices. How these questions are answered will vary from one political system to the next. One major contribution that policy studies can make to the public debate over how far government should go in altering life-styles is to compare the similarities and differences in health promotion policy among nations.

Accounting for British and American Policy Choices

A comparison of health promotion policies in the United States and Great Britain recommends itself for two different sets of reasons. The first deals with the general utility of a comparative approach to the study of public policy; the second with the particular relevance of British and American comparative health policy experiences. With regard to the first set of reasons, I would suggest that public policy is best studied and understood in a comparative framework because policymakers themselves frequently look to other countries for policy inspiration, ideals, models, and support.[33] The reasons for this comparative approach to policy-making include simple expediency and convenience, the need to buttress political arguments through reference to the examples of others, and the related desire to emulate proven policy successes and avoid demonstrated failures.

The potential benefits of a comparative policy perspective are enhanced in the case of Anglo-American studies by the fact that each country, despite significant differences in health delivery systems, demographic size and structure, and political architecture, views the other's policy experiences as particularly relevant to its own. There are a number of cultural and political reasons for this. First and foremost is the mutual accessibility of these experiences due to a common language. The ability to communicate easily has allowed policymakers in each country to draw upon the experiences and expertise of the other through newspaper accounts, the mass media, bureaucratic fact-finding trips, and direct legislative contacts.

The sharing of a common language is reinforced by a common political heritage, and especially the importance that individual well-being plays in the democratic value system. Bressler's comment, noted earlier, that one of the "master-values" of a democracy is the emphasis on the importance of human life and well-being applies equally to the United States and Great Britain.

A third explanation for the diffusion of policy ideas and experiences between the United States and Great Britain and, therefore, the utility of a comparative study of the two, lies in the fact that each, at different times and in different areas, has been viewed as a health policy leader or innovator. Richard Shyrock, for many years one of the foremost students of public health in the United States, noted some time ago the singular impact of British public health policy on the United States during the nine-

[33] Howard M. Leichter, *A Comparative Approach to Policy Analysis* (New York: Cambridge University Press, 1979).

teenth century, when the British were developing "the most effective local and national sanitary institutions which modern society had yet devised." Describing the various phases of sanitary reform, Shyrock noted, "All these phases of British activity made a marked impression upon contemporary American observers; to such an extent, indeed, that the British health movement operated as one of the chief causes of subsequent American sanitary reform."[34]

Today, however, the flow of policy information and inspiration is often in the other direction. Explaining and lamenting recent British priggishness concerning both smoking and drinking, Bernard Levin, a columnist for the London *Times*, recently noted that "now in these matters, the rule is: what America does today, Britain does tomorrow."[35] There have been times when the impact of one nation's experience on the other has been direct and unadulterated. This was the case with the adoption in 1966 of health warnings on cigarette packages in Britain. According to Laurie Pavitt, M.P. (Labor), who proposed the warning label, "I got the idea from talking to the U.S. Surgeon General and simply translated it into Westminster jargon."[36] It is difficult to read a British newspaper account, government report, court decision, or Parliamentary statement on health promotion in which the standard of comparison is not the United States.

Finally, Anglo-American comparative health policy analysis is justified if for no other reason than that the two health care systems have long held a peculiar fascination for, and attracted considerable attention in, each other's country. This is in part due to the fact that the two countries have taken significantly different approaches to providing health care for their citizens. Britain has the only Western democratic health-care system based upon the principle of comprehensive care for the entire population, divorced from ability to pay, and free at the point of delivery. For those Americans familiar with it, the British system of "socialized medicine" leaves few neutral in their opinions.

Despite discontent with problems of long queues for surgery and other medical care, the British remain proud of their health care system and see the U.S. system as uncaring and extraordinarily costly. One recent report in a British publication noted that "health-mad Americans are busily tak-

[34] Richard H. Shyrock, "The Origins and Significance of the Public Health Movement in the United States," *Annals of Medical History* 1 (January 1929): 646.

[35] *Times* (London), 28 January 1988. For other examples of the omnipresence of the American standard see Geoffrey Cannon, "The Food Scandal," *Times* (London), 13 June 1984; "Opinion of Lord Jauncy" *in causa* Mrs. Catherine McColl against Strathclyde Regional Council, 1984 (the case dealt with fluoridation); and Keith Tones, "On the Road to Health," *Health and Social Service Journal* (3 February 1983): 136.

[36] Rodney Deitch, "Talking Politics: A Man With a Mission," *World Medicine* (18 September 1982): 55.

ing their own blood pressures and checking diets. But when self-help preventive medicine fails they have to fall back on a health care system that can impose crippling burdens on the ill." The writer went on to note that the U.S. system relies predominantly on private health insurance that typically does not cover the full range of health care services. "So the American public is not getting the comprehensive service we expect in Britain. However, they appear to have been so brainwashed about 'socialised medicine' that the majority appear not to realise what a bad deal they are getting."[37]

The love-hate relationship that has often characterized British attitudes toward the U.S. health care system is nicely summarized by reaction to a recent plan by the Thatcher Government to reform the NHS by creating greater choice and competition within the system. Commenting on the proposed reform, the *Economist* editorialized that "some of the thinking behind these ideas is based on American experience. That sounds worrying, since among rich countries America has the most expensive and least efficient health care. But recent American studies have shown that competition, if properly regulated, can raise the quality of health care and may reduce its cost."[38]

Dr. David Owen, leader of the British Social Democratic party, was less ambivalent in his reaction: "The commercialization of health care is the primrose path down which inexorably lies American medicine: first-rate treatment for the wealthy and 10th-rate treatment for the poor."[39] Whatever the opinion, the standard of comparison remains the United States.

Whether it be to assess the differential impact of political structure and culture, or simply to gain insights into the policy approaches and perspectives of a nation for which we have a particular affinity and upon which we have often relied for inspiration, there is much to be learned from a comparison of British and American health promotion policies.

An Accounting Scheme

Despite the mutual accessibility of their policy experiences, and the sharing of a common democratic and postindustrial ethos, British and American policymakers have neither necessarily followed the same policy paths nor framed the life-style and health debate in precisely the same fashion. To understand and explain how and why they have differed requires a comparative analytical framework. I have developed such a scheme else-

[37] Potter, "Showdown," 1140.

[38] *Economist*, 4 February 1989, 15.

[39] *New York Times*, 1 February 1989.

where, building upon the work of Robert Alford.[40] I have suggested that to bring some semblance of order to the enormous number of policy-relevant variables that might explain similarities and differences in public policy among political systems and, therefore, to make the comparative study of public policy more manageable, it is necessary to have an accounting or organizing scheme. Such a scheme would enable us to proceed systematically in trying to understand and explain why different governments do what they do. It would allow us to inventory those factors that influence the setting of the policy agenda, the processing of policy demands, and the selection of policy options in different political settings. The accounting scheme I developed identified four categories of factors or variables that influence the policy process in any political system. These are situational, structural, cultural, and environmental variables. I will elaborate upon each, using examples from the area of health promotion policy.

SITUATIONAL FACTORS

Situational factors are more or less transient, impermanent, or idiosyncratic conditions or events that cause governments to consider or actually take public action. Situational factors include dramatic events such as natural disasters, epidemics, changes in government or political leadership, technological or medical breakthroughs, and economic cycles, or seemingly less profound events like the death of a movie actor from AIDS, the publication of a government report on a public health problem, and even chance meetings between public officials of two nations. Thus, for example, as noted earlier, a meeting between a British M.P. and an American surgeon general led to the adoption of warning labels on cigarette packages in Britain similar to those already in effect in the United States.

STRUCTURAL FACTORS

In the absence of facilitating institutional arrangements, situational factors are unlikely to lead to dramatic policy departures by themselves. Precisely how, or indeed if, a particular event or experience is transformed into a policy proposal is determined by the structural context in which it occurs. These factors have a more sustained and therefore generally more predictable impact on policy than situational factors. Structural factors are the essentially permanent and persistent features of a nation's society, polity, and economy. Included in this category are such variables as economic factors, including type of economic system and level of economic

[40] See Leichter, *A Comparative Approach*; and Robert Alford, *Bureaucracy and Participation: Political Culture in Four Wisconsin Cities* (Chicago: Rand McNally, 1969), 2.

development; political factors, including political organization, form of government, and the nature or style of the political and policy process; and social factors, including the demographic structure and epidemiological profile of a nation.

Of particular interest and relevance to this book, and the comparative study of British and American public policy in general, is the impact of political structure on public policy. An issue of long-standing interest to political scientists has been the comparative impact of formal political arrangements on public policy. Because of their obvious differences in political structure (separated versus fused powers, responsible versus undisciplined parties, and federal versus unitary governments), comparisons of British and American politics and policies are particularly promising in revealing the impact of structure on policy. Several years ago, for example, J. Roland Pennock examined the impact of "responsible government" and separated powers on agricultural subsidies and price support policies in the two countries. Pennock was particularly interested in the question of agricultural interest group access in a parliamentary versus a presidential system. He concluded that contrary to conventional wisdom, British-style "responsible government" was not "markedly superior to the [American] system of separated powers in its ability to resist the demands of special interests. . . . From [the] evidence it appears that in Britain, the all-or-none nature of party competition may make leaders extremely sensitive to the demands of pressure groups, and party discipline may be used to suppress elements in the party that would like to resist the demands of those groups."[41]

Cynthia Enloe, looking at a different facet of political structure, the distribution of political power in the United States and Britain (as well as the Soviet Union and Japan), found that in pollution control policy, differences in political structure do make a difference in public policy outcomes. "Britain's centralized political system has allowed it to engage in long-range planning to an extent unknown in more fragmented systems such as the United States."[42]

Hence, the evidence on the impact of political structure on public policy seems unclear. To the question, "Does political structure make a difference in the course, content, and probable outcome of health promotion policies in Britain and America?" I would answer "yes, but not in a consistent and predetermined fashion."

Another structural variable given attention in this book is the difference in political "style," or the way these two political systems process

[41] J. Roland Pennock, " 'Responsible Government,' Separated Powers, and Special Interests: Agricultural Subsidies in Britain and America," *The American Political Science Review* 56 (September 1962): 631.

[42] Cynthia Enloe, *The Politics of Pollution in a Comparative Perspective* (New York: David McKay Company, 1975), 314.

policy issues. A number of scholars, including Samuel Beer, Dennis Kavanagh, and David Vogel, have referred to the "consensual" and "corporatist" nature of British politics and policy-making.[43] In a recent study of environmental policy in the two countries, David Vogel sketched two different regulatory policy styles. He found that "regulation of industry tends to be more informal in Britain than in the United States, more flexible and more private."[44] In addition, there are fewer access points in the regulatory process in Britain, and there is a strong preference for voluntary agreements and restraints compared to the more common reliance on mandatory controls in the United States. One result is that British policy allegedly changes or responds more slowly and incrementally, as consensus between government and industry is arduously sought and conflict studiously avoided. Because regulatory policy is formulated and implemented in a more private, structured, gradual, cooperative, and flexible environment, there tends to be less overt political conflict over policy in Britain than the United States.

Closely related to the consensual bias in British policy-making has been a "corporatist" inclination. In the context of British and Western European politics, "corporatism" refers to a style of policy-making in which there is close consultation and collaboration among interest groups (especially labor and business groups) and government. Since 1945 British Labor (for example, under Harold Wilson) and Conservative (for example, under Harold Macmillan and Edward Heath) governments have periodically experimented with a corporatist approach to economic policy-making.[45] Although British corporatism was never as pronounced or successful as that of other Western European countries like the Netherlands, Austria, Denmark, West Germany, or Sweden, and although it certainly has fallen out of favor under the Thatcher government, facets of it remain, if only in the form of accepting the notion of consultation and cooperation between interest groups and the government.[46]

[43] See, for example, Samuel H. Beer, *Britain against Itself: The Political Contradictions of Collectivism* (New York: Norton, 1982); Dennis Kavanagh, *Thatcherism and British Politics: The End of Consensus?* (New York: Oxford University Press, 1987); David Vogel, *National Styles of Regulation: Environmental Policy in Great Britain and the United States* (Ithaca: Cornell University Press, 1986), 220.

[44] Vogel, *National Styles*, 220.

[45] See, for example, Andrew Shonfield, *The Use of Public Power*, ed. and intro. Zuzanna Shonfield (Oxford: Oxford University Press, 1982), 109–15; Andrew Shonfield, *Modern Capitalism* (New York: Oxford University Press, 1965), 161–63; Keith Middlemas, *The Politics of Industrial Society* (London: Andre Deutsch, 1979); and Paul Addison, "The Road From 1945," in *Ruling Performance: British Government from Attlee to Thatcher* (Oxford: Basil Blackwell, 1987), 16–17.

[46] Ian Budge et al., *The Changing British Political System*, 2d ed. (London: Longman, 1988), 36–38.

The question that remains, and one to which I will return throughout the book, is whether noncoercive, voluntary, consensual, or quasi-corporatist policy-making is more likely to minimize conflict and lead to effective resolution of public policy issues than the regulatory approach used in the United States. My quick answer at this point is not necessarily. Indeed, I will demonstrate that under certain circumstances such a seemingly benign approach may perpetuate and exacerbate conflict and inhibit successful policy solutions.

CULTURAL FACTORS

Political style is closely related to the third fundamental variable affecting the policy process, a nation's political and social culture. Indeed, political style grows out of and reflects a nation's social and political values. Cultural factors are the values, norms, and ideals of a society. They include the rules of the game that govern and make predictable social, economic, and political relationships. They include those ideals that society deems to be precious. One important difference between British and American social values involves the role of religion in each society. In comparing British and American attitudes toward the "New Right" agenda in each country, Samuel Brittan notes the difficulty in exact comparisons because, "Religious affiliation is very much less, and there is an absence [in Britain] of the vociferous fundamentalism so striking in the USA."[47] The prominence of religion, especially in the American Bible Belt, has helped to invest life-style and health debates in this country with a great deal of moral content. Political culture, a subcategory of the general culture, refers to those ideals, beliefs, expectations, and attitudes concerning the appropriate role and processes of government, and the nature of the relationship among individuals, groups, and the state. One important feature of the British political culture has been the willingness on the part of the working-class majority to defer to rule by their social and economic "betters." Over a century ago Walter Bagehot, in his classic study *The English Constitution*, referred to England as a "deferential nation." According to Bagehot, the British accepted the notion that "certain persons are by common consent agreed to be wiser than others, and their opinion is, by consent, to rank for much more than its numerical value."[48] Some persons are, in other words, born to govern. This notion, along with the reciprocal concept of noblesse oblige, has historically resulted in greater willingness to allow a benevolent government to act on behalf of the common-

[47] Samuel Brittan, *A Restatement of Economic Liberalism* (Atlantic Highlands, New Jersey: Humanities Press International, 1988), 243.
[48] Quoted in Philip Norton, *The British Polity* (New York: Longman, 1984), 29.

weal. The people defer to those in government with the expectation that they can be trusted to "do the right thing."[49] This point was raised in the course of the debate over mandatory seat-belt use when some opponents argued that the law would be virtually unenforceable and highly unpopular. Proponents countered that the British were a law-abiding people and the measure would be virtually self-enforcing. And it was. The compliance rate has been consistently around 95 percent, compared to about 40 to 50 percent in most American states with compulsory seat-belt laws.

The cultural propensity to trust government is related to another deep-rooted facet of the British political value system, collectivism. According to A. V. Dicey, in his classic study *Law and Opinion in England*, "By collectivism is here meant the school of opinion . . . which favors the intervention of the State, even at some sacrifice of individual freedom, for the purpose of conferring benefit upon the masses of the people."[50] Although consensus, deference, and collectivism have all suffered erosion in the last decade and are clearly not as strong as they were in Bagehot's and Dicey's time, each continues to influence the British policy process.

Americans, by way of contrast, tend to be highly suspicious of government and resist the notion of a benevolent state. This point is made convincingly by Anthony King, who compared public policy differences, and particularly variation in the scope of governmental activity, among five Western nations: the United States, Canada, Britain, France, and West Germany. King examined two broad policy areas: public versus private ownership of various industries, and the provision of social services. His major conclusion was that in every policy area except one—education—the state has played a far less active role in the United States than in any of the other nations. How does one account for this difference? After dismissing explanations involving differences in elite roles, mass behavior, interest group activity, and political institutions, King concluded that "the state plays a more limited role in America than elsewhere because Americans, more than other people, want it to play a more limited role. In other words, the most satisfactory explanation is one in terms of Americans' beliefs and assumptions, especially their beliefs and assumptions about government."[51] Throughout this book I will show how the abhorrence of Big Brother constantly influences debates over government-sponsored health promotion policies in the United States. Whereas the British

[49] See Samuel H. Beer, *British Politics in the Collectivist Age* (New York: Alfred A. Knopf, 1965), 3–5.

[50] A. V. Dicey, *Lectures on the Relation between Law and Public Opinion in England during the Nineteenth Century* (London: Macmillan, first published 1905, quoted in 1952 edition), 64.

[51] Anthony King, "Ideas, Institutions and the Policies of Governments: A Comparative Analysis," part 3, *British Journal of Political Science* (October 1973): 418.

have been willing to accept even an unpopular law such as mandatory seat belts, many Americans, leery of state interference in their personal lives, tend to view such laws as opportunities to subvert the will of government.

Despite the far less sympathetic attitude toward government involvement in the United States than in Britain, ironically, it is in the former that the state has gone farther in promoting health through public policy. A major reason is that there are significant differences in general social values between the two countries that have, in a sense, diluted the impact of political culture. I will show, for instance, that the British have a far more tolerant attitude toward alcohol consumption than Americans. Hence, the British have been less willing to interfere with social drinking than the Americans, who are currently going through one of their periodic bouts of puritanism.

I should note here that although it is possible to identify a predominant set of political and social values that distinguish one nation from another and affect the course and content of policymaking in each, it does not follow that all persons or leaders in a society subscribe to these values. Nor is it the case that fundamental belief systems are unchanging. A number of observers have wondered, for example, if the Thatcher years represent an end to consensus and collectivism in British politics.[52] Samuel Beer, in the sequel to his classic study *British Politics in the Collectivist Age*, writes about the British "cultural revolution" and particularly the "collapse of deference" that he believes began in the mid-1960s. Similarly Dennis Kavanagh, despite the tentativeness of the subtitle of his recent book *Thatcherism and British Politics: The End of Consensus?*, sees the fundamental British political value of consensus, breaking down. Although it may be premature to speculate on the long-term impact of Thatcherism on British social and political values, there is little doubt that Thatcher's own belief system diverges significantly from the noblesse oblige, collectivist, and consensual values that have characterized British political culture since the latter part of the nineteenth century. Particularly noteworthy from the perspective of this book is Thatcher's emphasis on encouraging an informed, personally responsible, and self-reliant public to make the "right" economic and social choices rather than rely on a benign, paternalistic government to impose those choices. In this sense Thatcher's values are reminiscent of the Tory belief system that characterized the early parts of the nineteenth century. According to Dicey, Toryism or "legislative quiescence" was the dominant current of public

[52] See, for example, Kavanagh, *Thatcherism*; and Raymond Plant, "The Resurgence of Ideology," in *Developments in British Politics*, ed. Henry Drucker, Patrick Dunleavy, Andrew Gamble, and Gillian Peele (New York: St. Martin's, 1983), 7–29.

opinion (that is, political culture) from 1800 to 1830. It was subsequently replaced by Benthamism or "individualism," from 1825 to 1870, and then collectivism from 1865 onward.[53] Dicey's characterization of "main currents of public opinion" alerts us to the fact that dominant political values can change.

Whatever the impact of Thatcherism, or Reaganism, for that matter, on British or American health promotion policy, it is clear that social and political culture play a decisive role in defining, processing, and selecting policy problems and solutions. In the context of health promotion policies, it is my judgment that neither Thatcher's nor Reagan's personal political or social values or agendas have had much of an impact on health policy. In part this is because such policies have been of only peripheral interest to either leader, and in part because other political or social factors proved more important and resilient than either Thatcher's or Reagan's own views on these issues.

ENVIRONMENTAL FACTORS

The fourth and final variable in my accounting scheme is environmental factors: events, structures, and values that exist outside the boundaries of a political system but influence decisions within that system. Many public health problems, including contagious and infectious diseases such as, AIDS or influenza, often emanate from outside one system but require responsive and protective policies, such as inoculation, testing, or quarantine, in another. In addition, there are many examples of decisions made by transnational organizations such as the World Health Organization or, in the case of Britain, the European Economic Community (EEC), that influence national policy choices. Recently the EEC has, for example, issued or proposed regulations dealing with the size of health warnings on cigarette packages, limitations on smoking in public places, community harmonization of duties on alcoholic beverages, and various safety features on automobiles. These regulations have been especially irksome to Thatcher, who sees many of the EEC proposals as "socialism through the back door." In the years ahead, as the *EEC* becomes more integrated, one can expect it to be the source of a number of life-style-related public policies in Britain.

The most important environmental factor identified in this study is that of policy diffusion. In general, policymakers rarely embark upon entirely new courses of action; they borrow heavily from an apparently finite, existing repertoire of policy solutions. They tend, when faced with a particular problem, to look for an analogous situation in another system,

[53] See Dicey, *Lectures*, 62–65.

either to emulate the solutions or to avoid the mistakes of others. I argued earlier in this chapter that there is a long and especially strong tradition of Anglo-American mutual awareness of common public health problems and policy options. Over the last few decades the British have been particularly attentive to health promotion policy developments in the United States, if only in some instances as a reminder of how not to proceed.

Finally, there are some areas of public health policy that lend themselves more readily to policy diffusion than others. In the case of AIDS, for example, which as of January 1990 had been the subject of five international conferences, sharing of policy as well as strictly technical information has become virtually institutionalized. As in the case of situational factors, however, policy diffusion is often less a cause for following a general policy approach than an occasion for considering a particular policy option.

In sum, variations in the content and direction of British and American health promotion policy can be accounted for in terms of a variety of situational, structural, cultural, and environmental factors. Rarely can the variations in the policies among these nations be explained by any single factor. Nevertheless, in this book I will show that both social and political culture and political structure account for most of the differences in health promotion policy in these two countries.

There has been a dramatic national and international shift in the field of health policy. In the United States this shift has affected virtually every adult American, touching upon his or her drinking, driving, smoking, nutritional, and sexual habits. Across the nation governments at various levels have banned "happy hours," increased the legal alcohol-drinking age, adopted mandatory seat-belt laws, and introduced drug and AIDS testing. This "second public health revolution" has affected people in every conceivable jurisdiction and circumstance. The U.S. Army and the Holden, Massachusetts, police department have banned smoking in the workplace, and the city of Birmingham, Alabama, has encouraged employees to adopt good nutritional habits as part of its "Employee Wellness Program." There are probably few children in either country who have learned not only to "Say No to Drugs," but about the virtues of condoms as well.

Public policymakers in the United States and Great Britain, of all political persuasions, have accepted the new-perspective assertion that imprudent life-style choices, along with environmental and workplace hazards, are among the most important factors contributing to the morbidity and premature mortality of the British and American people. There is near-universal agreement that people should wear seat belts, not smoke, drink moderately (and rarely, if ever, before driving or during pregnancy), eat more nutritious foods, exercise regularly, and be more "responsible" in sexual behavior.

Where disagreement arises, and where some of the most important and fundamental issues of democratic theory and practice become involved, is over questions such as: How far, in a democratic society, the state can go in regulating private behavior? When are people free to be foolish? When do the legitimate police powers of the state become either excessive paternalism or its more malevolent cousin Big Brotherism? Is there danger in an area such as health policy, filled as it is with good intentions and causes, of starting down a slippery slope of state intervention from which there is no logical exit? Where do we draw the line between private and individual responsibility, and public and collective responsibility in health promotion and disease prevention? These questions all touch upon one of the most enduring and critical political conflicts faced in any democratic society. As a British parliamentary committee noted over fifty years ago, "Between liberty and government there is an age-long conflict."[54] In the course of this book I will examine this most fundamental of all political conflicts as it applies to specific cases of life-style modification policy, including the consumption of tobacco and alcohol, motor-vehicle safety, and the prevention of AIDS. My purpose is to compare and explain the policy choices in Britain and America as each responded to these urgent public health problems.

I begin, in Chapter 2, by examining the first public health revolution of the mid-nineteenth century to early twentieth century. This historical excursion is recommended for two reasons. First, current problems and policy responses are rarely divorced from prior policy experiences, problems, or routines, and they often trace their origins, inspiration, and justification back to earlier times. Second, this historical approach allows me to underscore the point that public health problems and policy responses have an existential base. The first public health revolution, much like the current one, was in response to dramatic societal changes, and it too raised the specter of securing improved health at the cost of limiting personal freedom. In Chapter 3 I will discuss the origins, issues, and implications of the second public health revolution. This chapter provides a general preview of the issues raised in the specific areas of health promotion policy discussed in Chapters 4 through 7. These chapters deal in turn with comparisons of the dimensions of the public health problem, the

[54] Finance and Industry, Royal Commission Report 1931–32 vol. xiii), Cmd. 3897, pp. 4–5, quoted in W. H. Greenleaf, *The British Political Tradition*, vol. 1 (London: Methuen, 1983), 31.

major issues, and the evolution of policy responses associated with smoking, alcohol abuse, road safety, and AIDS. In each chapter I will emphasize the tensions of the collective and individual need to maximize both public health and personal freedom. I selected these issues not only because they involve some of the major sources of premature death in both countries, but also they reflect the conflict between the desire to promote responsible and healthy behavior and protect individual freedom. Finally, in Chapter 8 I review some of the major findings of the book and suggest my own answers, using the evaluative framework developed in this chapter, to the questions of when and under what circumstances is it appropriate for the state to intervene in life-style decisions.

Two

The Health of Nations: The First Public Health Revolution

Rarely in history has public health policy had as its sole purpose the promotion of good health or the prevention of disease. As often as not, it has sought to secure other critical social and political values, including moral renewal, reduction of poverty, improved economic efficiency, and strengthening of national defense. Many of these concerns were especially evident in the nineteenth century, when nations such as the United States and Great Britain were forced to respond to the social and political traumas associated with industrialization and urbanization. The result was the first public health revolution, during which industrializing nations took extraordinary public health measures.

In this chapter I will trace the development of early, that is, nineteenth- and early twentieth-century, British and American health promotion policies. I will focus on a sample of specific health concerns, cholera, smallpox, venereal disease, and infant mortality, to illustrate how these problems were portrayed as having implications beyond the personal suffering and tragedy associated with disease. It was by externalizing or socializing the implications of disease, especially the impact of ill health on their nations' wealth, physical vitality, and moral integrity, that British and American governments justified involving the state more intimately and frequently than ever before in the affairs of their citizens. Furthermore, it is my contention that the first public health revolution was in theory and substance the progenitor of the current public health revolution. Both flourished during times of widespread social change and frustration of medical science in the face of the most pressing health problems of the day.

In addition, in both periods social critics and policymakers confronted two faces of disease prevention and health promotion: collective and individual activities. In the nineteenth century, efforts to encourage or compel people to be vaccinated against smallpox proceeded along with the Chadwickian-inspired public sanitation movement. Similarly, in the twentieth century, the health promotion movement contained two, sometimes antagonistic, strategies: encouraging more responsible personal behavior, and eliminating the social and technological sources of ill health. Although the emphasis in this book is on the relationship between individual life-style and health, in this chapter and the next I will highlight some of the early efforts to improve the social and environmental incubators of disease.

I will begin with a discussion of Great Britain because it was there that the problems of industrialization were first manifested and the government first intervened aggressively in the name of health promotion. I will first describe the existential base of the public health revolution, and then how and why the state intervened to promote health and prevent disease.

Great Britain: The First Urban Nation

"If we look at the legislation of the last half-century [circa 1860–1915], it is safe to say that Public Health measures in one form or another have occupied a larger share of Parliamentary time than either trade, finance, education, or even national defence."[1] Whether or not this statement is statistically accurate, it certainly captures the apparent preoccupation of Victorian Britain with health issues. The question is, why was this the case? Why, beginning roughly in the third decade of the century and continuing for the next seventy-five years or so, did public health occupy so prominent a place on the British policy agenda?

At the outset it must be understood that in the context of nineteenth-century Britain there was nothing inevitable or predictable about the state promoting health. Ruth Hodgkinson reminds us that nineteenth-century Britain "had no recent experience of State paternalism or Enlightened Despotism. Therefore neither government nor the people felt strongly that the State was under any obligation to provide for the amelioration of health or living conditions. State intervention was unpopular for generations."[2] Government, in general, played only a minor role in the lives of nineteenth-century Britons. Yet despite the absence of a tradition of state paternalism, state intervention in health matters developed so rapidly and extensively that "it was not until the advent of compulsory education late in the century that the daily lives of Victorians were as widely affected by any government action as by state medicine."[3] The question remains, then: Why did state intervention in public health loom so large in nineteenth-century British life?

The most obvious, but far from complete, answer is that illness and

[1] William Brend, *Health and the State* (London: Constable and Company, 1917), 8.

[2] Ruth G. Hodgkinson, "Provision for Health and Welfare in England in the Nineteenth Century," *Proceedings of the XXIII International Congress of History of Medicine*, London, 2–9 September 1972: 176–96.

[3] Anthony S. Wohl, *Endangered Lives: Public Health in Victorian Britain* (Cambridge: Harvard University Press, 1983), 142.

disease loomed so large in a century of truly devastating epidemics: hundreds of thousands died between roughly 1830 and 1875 from cholera, smallpox, scarlet fever, measles, and typhus. To make matters worse, epidemic diseases tended to cluster: between 1836 and 1842 there were epidemics of influenza, typhus, smallpox, and scarlet fever, while from 1846 to 1849, typhus, typhoid, and cholera ravaged the nation.[4] In addition to epidemic diseases, thousands died each year from, or suffered other severe consequences of, such endemic diseases as tuberculosis, whooping cough, influenza, diphtheria, and dysentery. No health problem, however, was more notorious, tragic, or characteristic of the time than infant death. For much of the period under consideration there was virtually no improvement in the life chances of infants under one year of age (see Table 2-1).

Because these numbers are standardized they obscure the magnitude of the tragedy of infant death. For example, in 1861 alone, the infant death rate of 120 per 1,000 live births represented over 106,000 deaths—a number greater than any single year's epidemic disease toll. And, as F. B. Smith reminds us, these figures are almost certainly understated.[5] With all

TABLE 2-1
Infant Death Rate, England and Wales, 1839–1912 (Per 1,000 Lives)

Years	*Rate*	*Years*	*Rate*
1839-1840	153	1876-1880	144
1841-1845	147	1881-1885	139
1846-1850	161	1886-1890	145
1851-1855	156	1891-1895	151
1856-1860	150	1896-1900	156
1861-1865	151	1901-1905	138
1866-1870	157	1906-1910	117
1871-1875	153	1911-1912	113

Source: B. R. Mitchell and Phyllis Deane, *Abstract of British Historical Statistics* (Cambridge: Cambridge University Press, 1962), 36–37, cited in F. B. Smith, *The People's Health, 1830–1910* (New York: Holmes and Meier, 1979), 65.

[4] Bruce Haley, *The Healthy Body and Victorian Culture* (Cambridge: Harvard University Press, 1978), 6; and for a complete, if rather gruesome, chronicle of British epidemics, see Charles Creighton, *A History of Epidemics in Britain*, 2d ed. (New York: Barnes and Noble, 1965).

[5] See F. B. Smith, *The People's Health, 1830–1910* (New York: Holmes and Meier, 1979), 66.

this it is no wonder that "no topic more occupied the Victorian mind than Health—not religion, or politics, or Improvement, or Darwinism."[6] The prospect of ill health, disease, epidemics, and early death were a constant threat and concern of Victorian Britain. Why?

Although many of the health problems mentioned above predate industrialization, they were exacerbated and accelerated by it. Industrialization both requires and produces fundamental societal change. Among its primary requirements is a large and mobile labor force in concentrated urban centers. Typically, workers in early industrializing societies were recruited either from the rural populations within those nations or through immigration. The process of supplying an industrial workforce resulted in several societal changes, most obviously high population concentrations in industrial (including mining) towns and cities. In Great Britain these demographic shifts were unprecedented in scope and tempo. During the nineteenth century the British population residing in cities of 100,000 people or more increased from about one million in 1800 to 4.8 million in 1850, and 13.1 million in 1900. By mid-century Britain had become the world's first urban nation.[7] The 1851 British census reported that for the first time slightly over half of the population lived in towns or cities with a population of 10,000 persons or more; by the 1901 census the proportion had reached 77 percent.

The nature and magnitude of the social changes that accompanied nineteenth-century British urban growth were even more dramatic. To begin, housing conditions in the cities were deplorable. Overcrowded, rat-infested tenements could not accommodate the rapidly growing urban population. As a result, flimsy and unsafe structures were crowded even closer together, and all available "living" space, including hallways, attics, and cellars, was used.

Too many people crowded into too few houses produced problems involving unsanitary, inadequate, or nonexistent sewerage systems, pure water supplies, insect and vermin control, garbage disposal, air-pollution control, and burial procedures, as well as unsafe working conditions. These conditions hastened death and the spread of illness resulting from water-, air-, and food-borne diseases. Summarizing the attitude of many observers, Charles Kingsley, in his novel *Alton Locke*, lamented "these foul sties which civilisation rears—and calls Cities."[8]

Urban health problems were worsened by uninformed or misguided public policies. For example, to finance the late eighteenth-century Continental wars, Prime Minister William Pitt imposed a window tax that

[6] Haley, *The Healthy Body*, 3.

[7] Eric E. Lampard, "The Urbanizing World," in *The Victorian City*, ed. H. J. Dyos and Michael Wolff (London: Routledge and Kegan Paul, 1973), 1:4.

[8] Charles Kingsley, *Alton Locke* (London: Macmillan, 1893), 2:365.

remained in effect from 1784 until 1851. In order to avoid the tax, which was calculated on the total number of windows in a dwelling, many urban landlords bricked over windows, thus depriving renters of fresh air and sunlight.

There was, then, an existential basis for nineteenth-century British concern with disease and ill health. Industrialization and urbanization created and sustained unprecedented problems of social dislocation, environmental degradation, and sanitation that contributed to the introduction and spread of disease. Certainly, not all was bleak in Victorian towns and cities, since they continued to lure an ever-expanding population. Prospects for jobs, fortunes, social excitement, and marriage all appeared better in the urban than rural areas. But in moving from the countryside to the towns and cities, these new urbanites faced another, bleaker, prospect as well. Throughout most of the century there was a nearly perfect correlation between the size of the place of residence and the death rate; the more populous the place, the higher the death rate.

Advocating State Intervention

During the course of the nineteenth century, it became clear to many that the social and economic dislocations and human misery spawned by industrialization and urbanization required remedial action beyond that provided by private charities, existing poor laws, or "invisible hands." A positive exertion of state authority was needed. However, the concept of laissez-faire, which held sway in early nineteenth-century Britain, provided an ideological and practical obstacle to state intervention on behalf of the victims of industrialization. British policy would ultimately involve education, working conditions, labor laws, and social insurance, but it was in the area of public health that the state first began its assault on the consequences of industrialization.

That the earliest exertions of state authority should occur in public health was in part a function of the socially indiscriminate nature of health problems. Although inadequate housing and deplorable working conditions might offend the religious and moral sensibilities of the politically and socially powerful, they need not directly affect them. Epidemic diseases such as cholera, typhus, and smallpox, however, afflicted the high and low, with a differential impact, to be sure. The nation was reminded of how socially indiscriminate infectious diseases could be when in 1861 Queen Victoria's much beloved husband Prince Albert died of typhoid fever at the age of forty-two.

There was another reason for state quiescence to give way to state intervention in public health rather than in other problem areas: so many public health problems seemed to affect the young most severely. The need for government to do something about the tragedy of infant and child death, disease, and disfigurement was acknowledged even by the staunchest Tory or Liberal.

Beginning in the 1830s and 1840s, and continuing for the rest of the century, studies and commentary poured forth from social statisticians and reformers, epidemiologists, parliamentarians, physicians, clergymen, propagandists, and novelists documenting the nature, scale, and potential consequences of public health problems in Great Britain. Some of these studies were the product of prominent social reformers or investigators such as Edwin Chadwick, Sir John Simon, and Charles Booth. Other documentation took the form of studies by royal commissions and private organizations including the London, Manchester, and Birmingham Statistical Societies, the British Medical Association (BMA), the Ladies' Association for the Diffusion of Sanitary Knowledge, and scores of others. As Asa Briggs has noted, "By the end of Queen Victoria's reign, very much in English life had been measured."[9]

It is difficult to capture the extent to which the issue of health was kept before the public through novels, reviews, scholarly journals, occasional pamphlets, and especially newspapers and magazines. In fact, the widespread dissemination of information only became possible in the mid-nineteenth century with the development of a truly mass media. In 1801 there were a total of 129 magazines, newspapers, and reviews in England and Wales. By 1861 the number had increased to over 1,500, and by 1900 there were over 4,800.[10] In addition, remarkable as it may seem to us today, Victorians read, or at least purchased, government reports. Chadwick's Sanitary Report of 1842 had a circulation of over 10,000 copies! Never before had, or could have, the public, or indeed policymakers, been so well-informed about the nature, causes, and consequences of illness and disease.

The creation of a mass media prompted another important nineteenth-century political development, the emergence of public opinion. According to Samuel Beer, "In nineteenth-century politics the most striking change from the eighteenth century was probably the rise of public opinion."[11] In addition to the press, public opinion was often expressed in what Beer calls a "vast proliferation of voluntary associations." Not only were Victorians well-informed about their world, but they had developed the political mechanisms to articulate their concerns.

[9] Asa Briggs, "The Human Aggregate," in Dyos and Wolff, *The Victorian City*, 1:94.

[10] Michael Wolff and Celina Fox, "Pictures From Magazines," in Dyos and Wolff, *The Victorian City*, 2:575.

[11] Beer, *British Politics*, 43.

Major Health Issues

Given the multiplicity and diversity of individuals and organizations concerned, it is not surprising that they often reached different conclusions about the causes and remedies for the nation's health problems. Nevertheless, certain critical and general concerns and themes emerged, two of which loomed especially large. The first was the role of the state in encouraging or mandating more prudent behavior; the second, the collective origins and implications of disease and illness. Although the two were closely related, for analytical purposes it is useful to examine them separately. I will illustrate each by looking at specific public health problems. I will begin with the debate over smallpox vaccination because it spanned virtually the entire century and raised all of the questions I have identified as characteristic of health promotion policy debates.

MANDATING PRUDENCE: THE DEBATE OVER VACCINATION

Vaccination was the most radical, controversial, but ultimately successful public health measure of the nineteenth century. Probably no single measure, in any area of policy, brought citizens and government into more direct contact. One critic of vaccination wrote in 1879 that "the political question which it involves is also the largest, because it forces every family in the state to a compliance unlike anything else in the domain of politics."[12] Certainly no health policy prompted as sustained and hostile a debate. In 1906, more than a century after Edward Jenner first introduced vaccination into England, George Bernard Shaw launched an attack on the procedure that he called "nothing short of attempted murder."[13] In the course of this century-long debate, critics raised questions about the origins of disease, the reliability of the medical procedure, and whether or not the state had a right to force prudence upon its citizens.

Historically smallpox was one of the most infectious, disfiguring, dreaded, and dependable British health problems. From 1837, when national statistics were first recorded, until 1875, no fewer than 1300 persons died each year from the disease; as late as 1893, 1400 died from it. The disease tended, however, to visit in epidemic waves. Some of the most serious years were 1837–1840 (41,644 deaths), 1857–1859 (13,544 deaths), 1863–1865 (20,058 deaths), and 1870–1872 (42,084 deaths).[14]

Edward Jenner had introduced vaccination into England as early as

[12] "Vaccination the Largest Question of the Day," *Vaccination Tracts*, no. 14.

[13] Quoted in Bernard J. Stern, *Should We Be Vaccinated?* (New York: Harper & Brothers, 1927), 81.

[14] Creighton, *A History*, 614.

1798. Initially the government subsidized his efforts to produce a safe vaccine, but it did not formally sanction or require its use. It was not until the epidemic of 1837–1840 that the first Vaccination Act (1840) was adopted. The act, which provided free vaccination at public expense for anyone who wished it, was the first instance of state medicine in Britain, and an unprecedented assertion of state authority. Five times over the next seventy years (1853, 1867, 1871, 1898, and 1907), Parliament grappled with the medical and political implications of vaccination.

In 1853 Parliament made vaccination compulsory for all infants over three months old. Among the unintended consequences of the law was the founding of the Anti-Vaccination Society. Because of strong local opposition to the procedure, the 1853 act was only casually enforced. In 1867, however, the law was strengthened considerably. Local registrars were required to inform parents of their obligation to have children vaccinated within three months of birth, and to specify the time and place for the procedure. The law also allowed justices to require vaccination of any child under the age of fourteen. It was strengthened still further in 1871 with the introduction and aggressive administration of fines or imprisonment for parents who refused to have their children vaccinated. Parliament tried to overcome continued local reluctance to administer the law by requiring Local Government Boards to appoint vaccination officers.

The 1871 act represents both the high-water mark of compulsory vaccination in Britain and the acceleration of opposition to it. To complete the chronology of vaccination legislation before turning to the nature of the opposition, in 1898 Britain returned to permissive vaccination. A "conscience clause" was introduced into the law, allowing parents to avoid vaccination by testifying before a magistrate that they "believed" vaccination would harm their child. Penalties were abolished, and parents who wished to have their children vaccinated could have it done at home rather than in the hateful vaccination stations that had become a symbol of an oppressive governmental policy. Finally, in 1907 exemption was allowed merely by providing a written declaration of opposition to vaccination.

Opposition was restricted initially to the medical community, but by 1853, with the introduction of the compulsory provision, the issue became a full-scale national concern. Opponents, including George Bernard Shaw, Millicent Fawcett, and John Stuart Mill used the most damning language possible to condemn the practice and the government that first encouraged and then compelled its use. Opposition was expressed by religious organizations, national and local anti-vaccination associations, government Select Committees and Royal Commission hearings, public demonstrations, and a series of vaccination tracts published in the 1870s when it reached a climax. The titles of these tracts reflected both the substance and intensity of the opposition to vaccination. They included "Vaccination Laws: A Scandal to Public Honesty and Religion," "Vaccination: Evil in its Principles, False in its Reasons and Deadly in its Results," and "Compulsory Vaccination: A Desecration of Law, A Breaker of Homes and Persecutor of the Poor." In addition to the pamphlet war, there was also a nascent form of civil disobedience as many parents simply refused to have their children vaccinated. In Leicester, the seat of the anti-vaccination movement, over 1,100 parents were prosecuted in 1881 alone for failure to obey the law.

What was the basis for the highly charged opposition to vaccination? While there were many objections—one witness before the Royal Commission on Vaccination (1890–1892) gave twenty-six reasons why he opposed vaccination—three concerns dominated.[15] The first was the evidence concerning the medical reliability and efficacy of vaccination. Countless lay and scientific observers—the medical profession was itself divided over the issue—provided government committees and the press with the most gruesome anecdotes of vaccination gone awry. Many described disfiguration, ulceration, and fatalities from vaccination. Vaccination, it was charged, "propagates syphilis, consumption, and hereditary diseases," as well as "destroying the germs of the teeth during teething."[16]

In addition, many said the procedure simply did not work. "Vaccination was made compulsory in 1853, since which time small-pox has never been absent from the country, but has increased both in quantity and malignancy, each epidemic more severe and fatal than the one preceding."[17] There were more cases of smallpox in 1870–1872 than in 1840, and about 7 million more people as well, but successful refutations or explanations of the evidence had little impact on opponents.

A second major objection to vaccination was that it went against the laws of God. "It says to the people, No matter what your religious conviction as to the duty of protecting your offspring from the curse of vaccination, as you conceive it, no matter what your ideas of purity as the Gift of God may be, you must submit to our law. In thus saying and doing it stabs religion in its tenderest flesh."[18] Many lay religious leaders felt either that disease was the will of God and therefore not to be thwarted

[15] See *United Kingdom Fourth and Fifth Reports From the Royal Commission on Vaccination*, vol. 10 (1890–92): 289–91.

[16] "Compulsory Vaccination: A Desecration of Law, a Breaker of Homes and Persecutor of the Poor," *Vaccination Tracts*, no. 11, 3.

[17] "Facts and Figures," *Vaccination Tracts*, no. 2.

[18] "Vaccination the Largest Question of the Day," *Vaccination Tracts*, no. 14.

by man, or that vaccination, especially with cowpox vaccine, involved the introduction of an unnatural (i.e., sinful) substance into the human body.

Religious and medical concerns remained an important part of the debate throughout the century, but they were overshadowed after the 1860s by the issue of personal freedom. This particular line of opposition has a familiar ring to us today. Compulsory vaccination was described as invading in "the most odious, tyrannical and unexampled manner the liberty of the subject and the sanctity of home; [it] unspeakably degrades the freeborn citizen, not only depriving him of liberty of choice in a personal matter, but even denying him the possession of reason."[19]

Not only did compulsory vaccination immediately threaten personal freedom, but it represented a potential wedge in the door for future state incursions. "Moreover, compulsory vaccination if unrepealed is a gate by which tyranny after tyranny will enter the sanctuaries and sanctities of domestic life."[20] The metaphor has changed over the years, but the concern it symbolizes has persisted.

THE EXTERNALITIES OF DISEASE

The vaccination laws, enacted in response to a specific health problem, generated more passionate opposition than almost any other nineteenth-century public health measure because they involved matters of personal intimacy, freedom and morality. In addition, although proponents of vaccination did not make an explicitly collectivist argument, the vaccination laws must be seen in the context of the emerging view that personal ill health produced social costs (i.e., externalities). Health promotion policy, then, began to be justified in terms of saving money and enhancing the physical, economic, and moral vitality of the nation.

Among the public health themes to emerge during the nineteenth-century was that of the relationship between ill health and national wealth. There were several variations on this theme. One, used by reformers throughout the century, was the relationship among disease, poverty, and public spending on the poor, the "poor rates." The externalities of ill health were emphasized in 1839 by Thomas Southwood Smith, a sanitarian, physician, and member of the Poor Law Commission. According to Southwood Smith, "It is plain this disease [fever] is one of the main causes of the pressure on the poor rates. This pressure must continue and the same large amount of money spent year after year for the support of fam-

[19] George Gibbs, "Evils of Vaccination," in Stern, *Should We Be Vaccinated?*, 67–68.

[20] "Vaccination the Largest Question of the Day," 5.

ilies afflicted with fever, as long as the dreadful sources of fever which encompass the habitations of the poor are allowed to remain."[21]

The notion of the social origins of disease was enshrined in Edwin Chadwick's famous, and enormously influential, *Report on the Sanitary Condition of the Labouring Population of Great Britain* (1842). In the report Chadwick described how the deplorable living conditions of the poor led to disease and, often, the death of the breadwinner. This, in turn, reduced families to poverty and eventually led to the workhouse, where they became a burden on the ratepayers through poor relief. Furthermore, the poor were more likely to end up in fever hospitals, which also were publicly supported. Chadwick's conclusion, echoed decades later by another reformer, John Simon, was that disease prevention through improved sanitation and living conditions for the poor would be less costly than the treatment of disease or its consequences. According to Simon, "The prevention of evil, rather than the mitigation of the consequences of it, is not only the most beneficent, but the most economical course."[22] The notion that it is less costly to prevent disease than treat it has persisted virtually unchallenged—until recently—for centuries.

Edward Jenkins, a barrister and reformer, identified yet another social consequence of disease. Noting that "public health is public wealth," Jenkins saw the ill person as "a drain upon the resources of the state" not only in terms of direct support:

> One or more other persons in health are withdrawn from productive operations to expend their strength and time upon his [the ill person's] recovery. If one-third of a town, or city, or state, is suffering from disease, there is cast upon the other two-thirds a proportionately greater amount of exertion than would otherwise be required of them, and there is exacted from them a proportionately greater contribution to the general expenditure, while there is less capacity both of work and contribution in the whole community.[23]

Jenkins' argument that ill health directly and indirectly subtracted from the productive capacity of the nation was a particularly effective and long-lived tool in the hands of reformers. Toward the latter part of the century, and into the beginning of the twentieth century, British industrialists and politicians became deeply concerned about declining national efficiency due to the poor condition of the working class. Great Britain, they argued, was in danger of losing its preeminence in the fields of industry, commerce, and science to such countries as Germany and the

[21] Anthony S. Wohl, *Unfit For Human Habitation* (Cambridge: Harvard University Press, 1983), 146. Emphasis added.

[22] Quoted in ibid., 147.

[23] Alexander P. Stewart and Edward Jenkins, *The Medical and Legal Aspects of Sanitary Reform* (Leicester: Leicester University Press, 1969), 80.

United States. "At the center of the problem . . . lay the matter of physical efficiency. Great Britain was wasting her human resources."[24] B. S. Rowntree provided evidence of this in his study of York, entitled *Poverty: A Study of Town Life*. He reported that "a low standard of health prevails among the working classes. It therefore becomes obvious that the widespread existence of poverty in an industrial country like our own must seriously retard its development."[25] For a nation that thrived and prided itself on its leadership in industrialization, this was indeed a frightening prospect, and a legitimate basis for state intervention.

The danger that the poor physical condition of the working class posed for national economic productivity and efficiency was closely related to another, more ominous, consequence of disease and ill health. As early as the Crimean War (1853–1856), military authorities had expressed concern over the high rejection rate of working-class recruits because of their poor physical condition. Charles Kingsley warned at the time that "unless the physical deterioration of the lower classes is stopped by bold sanitary reform . . . we shall soon have rifles but no men to shoulder them."[26] It was not, however, until the Boer War (1899–1902) that the concern over the "physique of the nation" became a major national issue.

The Boer War should have been won quickly and easily. It was not. It took three years for the world's most powerful nation to defeat a ragtag army of Dutch farmers in what is now the Republic of South Africa. The explanation for this dismal military performance was the scandalously poor physical condition of working-class recruits. Military leaders complained of difficulty in finding enough healthy recruits to fill their ranks, and of the poor physical stamina and strength of many who were accepted for service. The issue was brought to a head in January 1902 with the publication of an article by Major General Sir John Frederick Maurice, entitled "Where to Get Men?" The general reported that within two years of seeking enlistment, three out of five potential recruits were either initially rejected or eventually dismissed because of physical infirmity. He suggested "that no nation was ever yet for any long time great and free, when the army it put in the field no longer represented its own virility and manhood."[27] General Maurice too attributed this frightful situation to

[24] Bentley B. Gilbert, *The Evolution of National Insurance in Great Britain* (London: Michael Joseph, 1966), 60.

[25] Quoted in J. R. Hay, *The Origins of Liberal Welfare Reforms, 1906–1914* (London: Macmillan, 1975), 54–55.

[26] Quoted in Wohl, *Endangered Lives*, 331.

[27] Quoted in United Kingdom Interdepartmental Committee on Physical Deterioration, *Report of the Inter-Departmental Committee on Physical Deterioration*, cd. 2175 (London: Her Majesty's Stationery Office, 1904), vol. 1, appendix 1.

the impoverished circumstances of the working class, from which the majority of the soldiers were recruited.

General Maurice's concern about the physical condition of the British soldier raised not only the prospect of a weakened military, but also that this "physical deterioration"—an oft-repeated phrase in the press—might go beyond working-class Army recruits. Could it be that all the British were deteriorating physically? The issue of the quality of the British race was raised not only by the military and the popular press, but also by eugenicists who were concerned about the increase in the "defective," or "inferior," or "degenerate stock" among the British population. Responding to "a deep interest . . . both in the lay and medical press" to General Maurice's writings, the Conservative government appointed an "Interdepartmental Committee on Physical Deterioration" to inquire into "the causes which have led to the rejection in recent years of so many recruits for the Army on the ground of physical disability."[28]

The Committee's report is interesting for what it reveals about contemporary notions of the causes of ill health and disease. At the outset it rejected the argument that there was increased physical deterioration among the British population. "It may be as well to state at once that the impressions gathered from the great majority of the witnesses examined do not support the belief that there is any general physical deterioration."[29] Yet it was still troubling to find that nearly one-third of the recruits were unfit for service. Why was this the case?

The answer was that the majority of the recruits came from the poorest of the working class; what one witness described as the "rubbish." In explaining the poor physical condition of this class, the committee echoed the argument of investigators since Chadwick of the social origins of disease; the urban environment, including overcrowding, pollution of the atmosphere, and poor working conditions, was largely to blame. But the committee found another reason as well, one which shifted part of the responsibility from environmental conditions to individual life-style. "On the other hand, in large classes of the community there has not been developed a desire for improvement commensurate with opportunities offered to them. Laziness, want of thrift, ignorance of household management, and particularly of the choice and preparation of food, filth, indifference to parental obligations, drunkenness, largely infect adults of both sexes, and press with terrible severity upon their children."[30]

More explicitly than anywhere else, the Physical Deterioration Report placed individual imprudence alongside social conditions as a cause of ill

[28] Ibid., 95.
[29] Ibid., 13.
[30] Ibid., 15.

health. Although the committee did not assess the relative impact of environmental and individual factors on the health of the nation, its recommendations included proposals to deal with both facets of the problem, with an emphasis on health education for the poor. Thus, the concept of individual responsibility for poor health through imprudent life-styles had been officially recognized. Personal imprudence posed a threat to the entire community through a weakened military force and perpetuation of poverty. Many eugenicists, like the great social reformer Sidney Webb, were convinced that "as a nation we are breeding largely from our inferior stock," and that public action was required if the nation was to survive.[31] Poverty itself was in no small part a function of personal irresponsibility. In a speech in 1906, David Lloyd George analyzed the "causes of poverty" and concluded that "the most fertile cause of all [was] a man's own improvident habits, such as drinking and gambling."[32] In sum, individual health and personal irresponsibility were now a collective concern.

A third theme related to the issue of the negative externalities of disease, couched initially in the idiom of "national defense," was that of the relationship between public health and private morality. The use of public legislation to enforce a code of morality became one of the most contentious and divisive issues in the latter decades of the century. One extraordinary point about this legislation is that much of it was introduced by Liberal governments that in most other policy areas displayed a visceral aversion to state meddling in the affairs of its citizens. Yet the moral revulsion prompted by antisocial practices such as prostitution and drunkenness allowed Liberal statesmen like Prime Minister William Gladstone to justify and endorse vigorous state involvement. Nineteenth-century Liberals—much like twentieth-century British and American conservatives—viewed the protection of public morality as a legitimate function of government. Not everyone, however, agreed. The public debate, which engaged religious and military leaders, feminists, and working-class reformers, as well as physicians, legislators, and the police, revolved around three laws that were euphemistically called the Contagious Diseases Acts. The first act was adopted in 1864, without fanfare or apparent public knowledge or interest, and sought to control the spread of venereal

[31] Sydney Webb, "Eugenics and the Poor Law—The Minority Report" (1909), 75. See also, for example, Major J. B. Paget, "The Health of the Nation," *English Review* 38 (March 1924): 377–85; Capt. E. Brown, "The Physique of the Nation," *English Review* 28 (March 1919): 242–50; and *Times* (London), 3 February 1908, 3; Dr. F. G. Shrubsall, "Health and Physique through the Centuries," *Nature*, no. 114 (November 1, 1924), 646–49.

[32] Quoted in Robert Eccleshall, *British Liberalism: Liberal Thought from the 1640s to 1980s* (London: Longman, 1986).

disease among British soldiers and sailors, a problem first documented by Florence Nightingale during the Crimean War.

The 1864 Contagious Diseases Act was limited to certain "protected districts" in ports and garrison towns in southern England and Ireland. It provided for the examination of any woman within a specified military district who, on the evidence submitted to a magistrate by a police officer, was suspected of being a "common prostitute" and having a venereal disease. The law, which was repealed by a subsequent act in 1866, went largely unenforced because Parliament did not provide funds for hospitals to administer examinations. The 1866 act, using basically the same evidential procedure, provided for the periodic and compulsory examination of known and suspected prostitutes living within a ten-mile radius of a military installation. This act, however, provided funds for hospitals, ominously called "lock" hospitals, to perform the examinations.

As a result of various subterfuges such as living outside the ten-mile district and commuting, or claiming to be menstruating when called for an examination, this law too proved ineffective.[33] In 1869, however, the act was strengthened to insure more effective and aggressive administration. Women could now be detained for up to nine months, during which time they were to receive religious and moral instruction. In addition, they could be detained until they were fit for examination, and the radius of the restricted areas was expanded.[34] In addition to these important substantive changes was an attitudinal change. Social reformers were able to provide data that showed a decline in juvenile delinquency, and the number of prostitutes and brothels, in the military areas covered by the acts. It was clear that the acts, initially intended to improve the efficiency of the military, had the potential of controlling both undesirable social elements and one factor contributing to the degeneration of the racial stock. "Extensionists," as they were called, formed the Association for the Extension of the Contagious Diseases Act and lobbied for application of the acts outside designated military areas.

The combination of the more aggressive administration and the effort to expand the application of the acts finally precipitated a massive movement against them. Between 1870 and 1885 the opposition, organized in a variety of preexisting or especially created associations, gathered over seventeen thousand petitions with over two and one-half million signatures, produced over five hundred books and pamphlets, held more than nine hundred public meetings, presented countless "memorials" to Parlia-

[33] See Smith, *The People's Health*, 120.

[34] Judith R. Walkowitz, *Prostitution and Victorian Society: Women, Class, and the State* (Cambridge: Cambridge University Press, 1980), 80.

ment, and attracted a host of prominent people to its cause, including the feminist Josephine Butler, John Stuart Mill, and Florence Nightingale.[35]

Opposition to the acts reflected a variety of sometimes incompatible religious, moral, and sociopolitical concerns. The major points of opposition were summarized in the *Report from the Select Committee on the Contagious Diseases Acts* (1882). While some expressed concern over the efficacy, enforceability, and reliability of the applied medical procedures, two objections overshadowed all others. First, opponents argued that the acts appeared to give official sanction to prostitution since they regulated rather than outlawed the practice. The acts, according to one witness before the Select Committee, were "a compact between the State and prostitution."[36] The State was not simply sanctioning "vicious sexual indulgence," it was facilitating it by removing or minimizing one of the main obstacles to it, fear of contracting venereal disease.

The second major objection to the acts was that they discriminated against women and deprived them of their constitutional rights. Women were subjected to compulsory medical examination and detention, without a trial or recourse to release by habeas corpus. In addition, men, who played an equal role in both the sexual deed and the spread of the venereal diseases, were free from the restraints, penalties, indignities, and pain (from "instrumental rape" during the examinations) to which women were subjected.

Despite the considerable effectiveness of the opposition, the Select Committee on the Contagious Diseases Acts recommended that the acts not be repealed. Nevertheless, opposition continued unabated, and in 1883 Parliament succumbed and amended the acts by repealing the provisions dealing with periodic compulsory examination and detention. Finally, in 1886, the notorious acts were repealed completely.

The Contagious Diseases Acts occupied an important place in the public health movement for nearly two decades. They contained all the features of what I have described as characteristic of most health promotion policies. They attempted to regulate highly intimate behavior and in the process limited personal freedom. The pattern of personal behavior involved was deemed morally obnoxious by the majority and was thought to have profound social implications. The Acts provide a classic example of a public health measure that quickly became as much a campaign over public morality as an effort at health promotion. Those familiar with the current debate over AIDS will find much in it that is familiar.

[35] Jeffrey Weeks, *Sex, Politics and Society* (London: Longman, 1981), 20.

[36] *United Kingdom Report from the Select Committee on the Contagious Disease Acts*, vol. 7 (1882): 220.

The Public Health Acts of 1848 and 1875

It is appropriate to conclude this discussion with a brief discussion of the Public Health Acts of 1848 and 1875 that frame the most intense period of nineteenth-century public health reform. The acts epitomize the essential abandonment of laissez-faire practices, if not principles, in the area of public health. The 1848 act heralded an era of aggressive state intervention in health promotion, and the 1875 act consolidated and reaffirmed that commitment. Despite their rather profound implications these acts, unlike those dealing with smallpox vaccination and venereal disease control, did not generate much popular passion. They simply did not pose the same threat to personal intimacy since they sought to regulate environmental conditions, not individual behavior. As S. E. Finer noted in his biography of Edwin Chadwick, one of the accomplishments of the Sanitary Report of 1842 was to shift the focus of public attention from individual failings to state responsibility in cleaning up the environment. According to Finer, "by proving that the individual had no control over these [unsanitary living] conditions the remedy was lifted at once from the regime of private initiative to that of public administration."[37]

The theme of the social origins of disease introduced by Chadwick would dominate much of nineteenth-century policy and thinking and would be revisited more than a century later in another British government study, the Report of the Working Group on Inequalities in Health (known as the Black Report). The Public Health Acts reflected the emerging consensus concerning the necessity of an active state role in disease prevention.

The Public Health Act of 1848 "was possibly the most significant piece of health legislation in the nineteenth century. It marked the first clear acceptance by the State of responsibility for the health of the people."[38] The architect of the act was Edwin Chadwick, a man who more than any other was responsible in these early years for documenting the extent of public health problems, relating them to environmental conditions, and spelling out some of their social implications. In his 1842 report Chadwick pointed out that the deplorable living conditions of the poor were an incubator not only of diseases that might affect the entire community, but also of potential working class radicalization and political instability. He warned that the urban slums harbored "a population that is young, inexperienced, ignorant, credulous, irritably passionate and dangerous."

[37] S. E. Finer, *The Life and Times of Sir Edwin Chadwick* (London: Methuen and Company, 1952), 217.

[38] Brian Watkin, *Documents on Health and Social Services: 1834 to the Present Day*, (London: Methuen and Company, 1975), 48.

In addition, Chadwick made what was to become a timeless health promotion policy claim; prevention of disease was cost-effective—in this case by reducing Poor Law rates.[39]

Chadwick's proposals did not find their way into legislation until 1848, when an anticipated European cholera epidemic helped insure passage of the Public Health Act. Although the act did not attract the widespread public attention and opposition that other health-related measures did, it was not without its detractors. One of these was the influential *Economist*, which at the time of the parliamentary debate over the act expressed a widely held view about the origins of disease, and the appropriate role of government: "Suffering and evil are nature's admonitions. They cannot be gotten rid of; and the impatient efforts of benevolence to banish them from the world by legislation . . . have always been more productive of evil than good."[40]

The act created a national General Board of Health that was to oversee the direction and implementation of public health and sanitation practices, including those dealing with drainage and sewerage, public health hazards and nuisances, location of burial sites and internment procedures, and handling of epidemics. In addition, the board was empowered to create local boards of health in any locality where either 10 percent of the ratepayers petitioned for one, or the annual death rate exceeded 23 per 1,000 people. Although the General Board of Health was ultimately to become a political casualty due to Chadwick's abrasive character and methods, the precedent the act set and the legacy it left, including the principle of state responsibility in disease prevention, marked the Public Health Act of 1848 as one of the truly extraordinary health achievements of the century.[41]

The Public Health Act of 1875 was the product of a Royal Sanitary Commission Report (1869–1871). The act was important less for any significant policy innovations than for its ratification and consolidation of the public health revolution of the previous three decades. The act realized two major goals. First, it consolidated the provisions of twenty-nine existing laws that covered virtually the entire range of state activity in public health policy. Second, the law clearly spelled out the duties and responsibilities of local health authorities in providing the proper environment for preventing disease and its spread. The significance of this act lies both in the breadth of the assumption of state interest and involve-

[39] Quoted in Knut Ringen, "Edwin Chadwick, the Market Ideology, and Sanitary Reform: On the Nature of the Nineteenth-Century Public Health Movement," *International Journal of Health Services* 9 (1979): 114.

[40] Quoted in John Gordon Freymann, "Medicine's Great Schism: Prevention vs. Cure: An Historical Interpretation," *Medical Care* 13 (July 1975): 527.

[41] See Watkin, *Documents*, 45–48; Ringen, "Edwin Chadwick," 113–18.

ment in public health, and in its testimony to how far that assumption had come in the course of the nineteenth century.

The United States: Coming of Age in an Urban World

There are a number of important similarities between the United States and Great Britain in their early public health problems, politics, and policies. Such similarities are predictable for two reasons. First, both shared common structural features such as industrial and urban societies, representative democracy with an expanding electorate, and capitalist economies supported by a laissez-faire philosophy. These conditions spawned similar health problems and structured similar policy responses. Second, as I noted in Chapter 1, British policy experiences exercised enormous influence on the United States. To give just one obvious example, John C. Griscom's *The Sanitary Condition of the Laboring Population of New York* (1845) bore considerable resemblance in methodology and conclusions, not to mention title, to Edwin Chadwick's *The Sanitary Condition of the Labouring Population of Great Britain* (1842). There were few health promotion problems or proposed policy responses in the United States that did not elicit some comparison with, or derive some lessons from, a similar British experience.

Although problems of infectious and communicable diseases and infant and maternal mortality loomed large in nineteenth-century America, they did not occupy the same place of prominence they did in Great Britain—and certainly not in the early part of the century. Indeed, as Charles Rosenberg points out in his study of the cholera epidemics of 1832, 1849, and 1866, "In 1832, most Americans regarded the United States as a land of health, virtue, and rustic simplicity."[42] The perception of a healthy America was partly a function of the fact that the United States was a more rural nation. For example, the percentage of Americans residing in cities of 20,000 people or more remained under 10 percent through 1850, while in Great Britain it had been nearly 17 percent as early as 1800, and 35 percent by 1850 (see Table 2-2).

The process of urbanization was, however, well on its way. In 1830, 8.8 percent of the U.S. population was classified as urban, by 1860, 19.8 percent, and by 1910, 45.6 percent. The urban population increased from about 200,000 people living in six cities in 1800 to over 18 million in 448 cities in 1890. This growth represented an 87-fold increase in the urban population during the century, while the total population increased 12-

[42] Charles E. Rosenberg, *The Cholera Years: The United States in 1832, 1849 and 1896* (Chicago: University of Chicago Press, 1968), 6.

TABLE 2-2
Percentage of Population in Cities of 20,000 People or More, United States and Great Britain, 1800-1900

Country	*1800*	*1850*	*1900*
United States	3.8	9.8	25.2
Great Britain	16.9	35.0	48.3

Source: *Encyclopedia of Social Reform* (Westport, Connecticut: Greenwood Press, 1970), 239.

fold.[43] In comparison to Great Britain, the image that emerges is of a later, sporadic process of urbanization in the United States, more geographically concentrated in the North Atlantic states, and externally fueled by immigration. But the consequences were much the same: in 1810 the crude death rate in New York City was 21 per 1,000; in 1857 it was 27 per 1,000. In addition, as in Great Britain, life chances in urban America were grimmer than in rural areas. In 1890, the national death rate among the rural population was 15.3 per 1,000, compared to 22.2 for the urban population and 24.6 in the metropolitan district of New York City.

And, as in Great Britain, there were reformers who analyzed the conditions, and consequences of, the poverty that was the concomitant of urbanization for so many. Griscom's study chronicled the "mournful and disgusting condition, in which many thousands of the subjects of our government pass their lives." Based upon his observations Griscom drew the following conclusions: "There is an immense amount of sickness, physical disability, and premature mortality, among the poorer classes"; "these are, to a large extent, unnecessary, being in a great degree the results of causes which are removable"; and "these physical evils are productive of moral evils of great magnitude and number, and which, if considered only in a pecuniary point of view should arouse the government and individuals to a consideration of the best means for their relief and prevention."[44]

Griscom's study was not officially undertaken and had little immediate policy impact; it would be over two decades before New York City created a Board of Health to implement an effective sanitary reform program. Nevertheless, it provided a model for much that was to follow in the field of public health.

Other cities, in other states, had their problems, albeit on a smaller

[43] See Adna F. Weber, *The Growth of Cities in the Nineteenth Century* (New York: Macmillan, 1899), reprinted in *The American City: A Documentary History*, ed. Charles N. Glaab (Homewood, Ill.: Dorsey Press, 1963), 181.

[44] George Rosen, *A History of Public Health* (New York: MD Publications, 1958), 238.

scale, and these situations too were the subject of popular and professional scrutiny. Contrary to the situation in Great Britain, however, there never developed the immense outpouring of government reports, pamphlets, and societies advocating public health reform in the United States. As Howard Kramer explains, "Nor did the periodical literature of the day abound in articles on slum conditions in American cities. Other reforms held the center of the stage. An imposing array of public questions—slavery, temperance, women's rights, prison reform, treatment of the insane, cruelty to animals—monopolized the nation's interest."[45]

It was in part because of these distractions that one of the century's most important public health documents, Lemuel Shattuck's Report of the Massachusetts Sanitary Commission (1850), went unimplemented. Shattuck argued that about 50 percent of all deaths were preventable. With little faith in the efficacy of medicine, Shattuck saw improved sanitary and environmental conditions as the major vehicle for preventing disease. He recommended the creation of a state board of health that could direct local authorities in appropriate sanitary measures. Like Griscom in New York, Shattuck had little immediate impact: the study was not even debated by the Massachusetts legislature. It was not until after the Civil War that the nation began implementing many of the recommendations of Griscom, Shattuck, and others.

Major Health Policy Issues

FROM LAISSEZ-FAIRE TO STATE INVOLVEMENT: THE REACTION TO CHOLERA

Few factors played as prominent a role in precipitating and justifying state involvement in disease prevention and health promotion as epidemics. And few epidemics held as much horror as cholera. It was not the disease's deadliness—other nineteenth-century diseases killed far more—but the swift and harsh way it claimed its victims. Individuals died, often within hours of the onset of the disease, from diarrhea that dehydrated them, leaving them "shrivelled like raisins with blackened extremities."[46]

Cholera first appeared in the United States in the early 1830s and last appeared in epidemic form in 1866. But the history of the three epidemics reflected the changing mood of the nation and its attitude toward the role of government in disease prevention. Charles Rosenberg, who has so marvelously captured that mood, demonstrates that the initial reaction to

[45] Howard D. Kramer, "The Beginnings of the Public Health Movement in the United States," *Bulletin of the History of Medicine* 21 (March–April 1947): 362.

[46] Smith, *The People's Health*, 230.

cholera was to view it as a manifestation of God's displeasure. According to the governor of New York, whose state was the hardest hit by the epidemic, "An infinitely wise and just God has seen fit to employ pestilence as one means of scourging the human race for their sins, and it seems an appropriate one for the sins of uncleanliness and intemperance."[47]

Irresponsible personal practices, then, could ignite God's wrath. Many in this predominantly rural country comforted themselves with the idea that such imprudence, and its consequences, were most often associated with urban living and the urban poor. The virtuous and hardworking, if not immune from cholera and other diseases, were at least less vulnerable. And, indeed, the urban poor, virtuous or not, *were* the most frequently afflicted. Overcrowded and unsanitary living conditions, inadequate housing, poor or nonexistent sewerage and safe water systems, all provided congenial hosts for water-, food-, and airborne diseases.

No government could merely sit back and rage against sin and proclaim that the afflicted were getting their just desserts. Towns and cities across the country set up cholera hospitals for the poor, directed the cleaning of streets, alleyways, open sewers, and canals, and established quarantines. But given the prevailing wisdom about the causes and distribution of disease, it is not surprising that these efforts were limited in scope and duration. They were not institutionalized to prevent future epidemics and promote good health, but rather dealt with the immediate problem at hand and then ceased. As long as sickness was largely equated with sin, government's role would be a limited one. This sentiment persisted, in the main, through the second cholera epidemic, in 1849. Although the medical profession increasingly recognized that environmental factors played an important role in the genesis and spread of disease, and was independent of vice or virtue, the dominant wisdom remained that "as a very general rule when a man gets sick it is his own fault, the result of either ignorance or presumption."[48]

Although only thirty-five years separated the first and last cholera epidemics, the demographic, political, and philosophical changes meanwhile in the nation were profound. First, the United States was a far more urban nation in 1866 than 1839—about one-fifth of the population in 1866 was urban compared to less than one-tenth in 1839. Second, American politicians and physicians were fully aware of the public health successes of the British. Dr. John Snow's demonstration that cholera was transmitted by contaminated water and could therefore be controlled by ensuring a

[47] Charles E. Rosenberg, "Social Class and Medical Care in Nineteenth-Century America: The Rise and Fall of the Dispensary," *Journal of the History of Medicine* 29 (January 1974): 41.

[48] *Halls Journal of Health*, vol. 1 (1854), cited in Rosenberg, *The Cholera Years*, 150.

clean water supply was published in 1855 and was well known and widely accepted on this side of the Atlantic by 1866.

Perhaps the most important policy response to the 1866 cholera epidemic, both substantively and symbolically, was the creation of the New York Metropolitan Board of Health, "a turning point in the history of public health not only in New York City, but in the United States as a whole."[49] The board, created just in time to deal with an anticipated cholera epidemic, was empowered to clean up the conditions that were now accepted as the environmental incubator of cholera: filthy streets, unsafe water and produce, and improper sewerage. In addition, the board required that all cases of cholera be immediately reported to the authorities. A sanitary team was then sent to disinfect the premises. The board also had the authority to establish quarantines and cholera hospitals. The metropolitan board was the first permanent, politically effective municipal health board in the country, and it proved an immediate success. Despite the doubling of New York City's population between 1839 and 1866, in the last epidemic the city suffered only about one-tenth the number of cholera deaths of the first. Most major cities shortly followed New York's lead.

The lessons from the changing perceptions of and reactions to cholera are important to our understanding of health promotion and disease prevention policies. By the last decades of the nineteenth-century, public health bureaucracies had become institutionalized. People, including large numbers of immigrants from countries like Prussia and Ireland, where the state was already more assertive, accepted and expected government to intervene to protect them from infectious and communicable diseases. Eventually the definition of this responsibility would involve ensuring safe water and milk supplies, trash collection and sewage disposal, healthy living and working environments and, in some localities, health dispensaries. Cholera was defeated by sanitarians, not physicians. The lesson that when medicine fails, public health, broadly defined, may provide the most promising protection against disease would become part of the health policy fabric of the United States. Finally, the cholera experience reminds us of the close identification in the minds of many between disease and life-style. Even as late as 1866, the *New York Times* still took the position that "cholera is especially the punishment of neglect of sani-

[49] Felix Marti-Ibanez, *History of American Medicine* (New York: MD Publications, 1959), 159. See also Charles E. Rosenberg, "The Bitter Fruit: Heredity, Disease, and Social Thought in Nineteenth-Century America," *Perspective on American History* 8 (1974): 193; Bert H. Brieger, "Sanitary Reform In New York City: Stephen Smith," in *Sickness and Health in America*, ed. Judith Walzer Leavitt and Ronald L. Numbers (Madison: University of Wisconsin Press, 1978), 359–73; and John Duffy, *A History of Public Health in New York City, 1866–1966* (New York: Russell Sage Foundation, 1974), 619.

tary laws; it is the curse of the dirty, the intemperate, and the degraded."[50] Although this view was becoming more equivocal, it was still held by many.

EXTERNALITIES: MORAL RECLAMATION, NATIONAL DEFENSE AND THE BATTLE AGAINST VENEREAL DISEASE

The relationship between sickness and sin was prominent and important in other nineteenth-century public health policy debates, especially those involving venereal disease. Periods of rapid social change often produce conservative reactions, as is happening in the United States and Britain today and occurred in the latter part of the nineteenth century and the early part of the twentieth century. Part of that earlier experience included the "social hygiene movement," or the "purity crusade." It involved social reformers of mixed ideals and goals: feminists, military and religious leaders, the medical community, and public health advocates.

Among the primary concerns of this diverse movement was the relationship between prostitution, which followed the trajectory of urbanization, and venereal disease. For the moral purists the relationship was clear; so too were the policy implications. "Sexual relations between men and women of easy virtue constitute the ultimate source of the overwhelming proportion of infections with syphilis and gonorrhea."[51] To stop the spread of venereal disease, prostitution must be stopped. Others, while agreeing that prostitution was a major factor in the spread of venereal disease, were more concerned about the implications of prostitution for public health than for public morality. Hence, one major division within the movement was between those favoring regulation of prostitution, and those who sought its complete abolition. The "regulationists," like their English cousins, believed that prostitution would always be around. The best way to deal with it was to restrict it to certain "red-light" districts and to control its noxious consequences through medical examinations of its practitioners. Some cities, including St. Louis, Cleveland, Detroit, Davenport, Iowa, Minneapolis, and San Francisco, adopted regulationist policies, with registration and medical examination of prostitutes. However, the controversy surrounding the British experience with the Contagious Diseases Acts, which was well known to Americans, and strong feminist and religious opposition, doomed the regulationist approach. A few cities attempted, unsuccessfully, to regulate prostitution into the twentieth century, but in the main these efforts were no longer central to the attack on prostitution by the 1890s.

[50] *New York Times*, 22 April 1866, quoted in Rosenberg, *The Cholera Years*, 218.

[51] John H. Stokes, *Today's World Problem in Disease Prevention* (Ottawa: F. A. Acland, 1923), 104.

The other approach, favored by most feminists and other reformers, involved a combination of education and progressive social policy. Public health education was needed to alert people to the moral and health implications of "deviate" sexual indulgences, those outside marriage, and to urge them to behave more responsibly. Enlightened social policies would help eradicate the conditions that forced women into prostitution, such as poor housing and few educational or job opportunities.

The campaign against prostitution and venereal disease nicely illustrates the point that health promotion policy is often an occasion for accomplishing other social and political objectives. From a public health perspective venereal disease, although not inconsequential, was less important than the attention it attracted. In 1900 the death rate from syphilis was 12.0 per 100,000 people; from typhoid, 31.3; from diphtheria, 40.3; from tuberculosis, 194.4; and from influenza or pneumonia, 202.2. In addition, as Ruth Rosen points out, it was probably the case "that the peak of women's engagement in prostitution took place between 1850 and 1900 rather than during the early years of the twentieth century, when, ironically, it [venereal disease] assumed the status of a major issue."[52] Prostitution and venereal disease became great issues because of their larger social and political implications, not because of the health risks they posed.

One of those implications was a by-product of the urban transformation described above. As the cluster of social and economic changes—urbanization, immigration, secularization, the breakdown of the extended family, and the changing status of women—proceeded, many saw the social and moral fabric of the nation endangered. The United States was a nation whose moral and social compass was spinning wildly. "This tension between personal morality and public order—a central aspect of Progressive ideology—lay at the heart of debate concerning venereal disease."[53]

Of particular concern was the perceived danger to the family. Social reformers, in arguments that have a remarkably contemporary sound, pointed to increases in divorce rates and the number of women working outside the home, and the decrease in family size as indicators of the precariousness of the American family. In 1911 former President Theodore Roosevelt warned white, middle-class Americans that if the low birth rate continued the nation was in danger of committing "race suicide."

Prostitution and venereal disease played an important role in the deterioration of the population, as well as in undermining the family. Men, in

[52] Ruth Rosen, *The Lost Sisterhood: Prostitution in America, 1900–1918* (Baltimore: Johns Hopkins University Press, 1982), 3.

[53] Allen M. Brandt, *No Magic Bullet: A Social History of Venereal Disease in the United States since 1880* (New York: Oxford University Press, 1985), 48.

self-indulgent and socially irresponsible fashion, were transmitting venereal diseases from harlots to innocent brides. The result was that "society paid for its neglect in wrecked homes, childless marriages, invalidism, blindness and insanity."[54] Prostitution and venereal disease, then, were both cause and manifestation of the nation's social, moral, and physical decline.

The response to the problem was twofold. First, states adopted laws that made proof of freedom from venereal disease a requirement for a marriage license. Michigan was the first state to do so, although it merely required a pledge from the groom that he was free of disease. In 1909, Washington became the first state to require a medical examination as proof, and other states quickly followed suit. The second approach was to require physicians to report cases of venereal disease to public health (or police) authorities. This was initially opposed by some in the medical community as a breach of doctor-patient confidentiality, but the majority of physicians, and most Progressive Era policymakers, accepted it as a necessary step in eradicating venereal disease. By 1922 all states had adopted mandatory reporting legislation.[55]

Both measures represented significant departures in acceptable medical and political practices concerning sex. Although some states had long required reporting cases of infectious diseases, venereal disease was clearly viewed as a different sort of health problem, a taboo in the minds of the general public and medical community. Venereal disease differed from other diseases in another way. While mid-nineteenth-century Americans saw cholera and other diseases partly in terms of divine retribution for personal irresponsibility, the connection was always somewhat oblique. In the case of syphilis and gonorrhea, however, the self-inflicted nature of the disease was clear and direct. According to Prince Morrow, a leading social hygienist, "Venereal disease seeks no man, it must be sought in order to be acquired."[56]

A second implication of venereal disease was that the threat it posed was not only to the nation's social fabric, but also to its national defense; venereal disease undermined the ability of the military to recruit a healthy fighting force. American recruiters during World War I had to reject as physically unfit a significant proportion of prospective recruits, especially because of venereal disease. In addition, there was concern that even those who were free of disease when they entered the armed forces would fall prey to temptation and contract venereal disease either at training centers in this country or when they went abroad. In 1917 Secretary of

[54] Stokes, *Today's World Problem*, 5.

[55] Mark Thomas Connelly, *The Response to Prostitution in the Progressive Era* (Chapel Hill: University of North Carolina Press, 1980), 79.

[56] Brandt, *No Magic Bullet*, 37.

the Navy Josephus Daniels urged a medical group to help the military deal with the problem because "we are fighting for the safety of democracy. Victory is jeopardized by the preventable diseases which destroy the fighting strength of armies and navies." He went on to say that "today as never before American manhood must be clean and fit. America stands in need of every ounce of strength." Finally, the secretary reminded his audience of the relationship between defense of the nation and protection of the family: "There lies upon us morally . . . the duty of leaving nothing undone to protect these young men from that contamination of their bodies which will not only impair their military efficiency but blast their lives for the future and return them to their homes a source of danger to their families and communities at large."[57]

The federal government responded to the immediate war needs by including in the Selective Service Act provisions to prohibit prostitution in areas adjacent to military training facilities. Wherever possible the military enlisted the assistance of local and state authorities in keeping the "boys clean." Where such assistance was either not forthcoming or inadequate, stronger measures were used to shut down red-light districts. Soldiers patrolled the areas, arrested prostitutes, and required them to take physical examinations. As Brandt has noted, these actions raised significant political and legal questions that transcended the current problem: "How far could a government go in the name of science and hygiene to abridge the rights of those who might threaten the greater good?"[58] This concern would remain at the heart of the health promotion debate throughout our century.

PUBLIC HEALTH AND PERSONAL FREEDOM: SMALLPOX VACCINATION

The question of how far government could go in the name of the common good was also at the heart of the American debate over compulsory smallpox vaccination. Although most of the arguments paralleled those heard in Great Britain, there was one major difference between the two countries. Except for a brief period (1813–1822), there has been no national vaccination policy in the United States. In 1813, Congress adopted "An Act to Encourage Vaccination," authorizing the President to appoint an agent to distribute vaccine through the post office system to any citizen requesting it. This act was repealed in 1822, and the issue, like most other public health measures, was left to each of the states.

Of the various policy concerns surrounding vaccination, the most con-

[57] Josephus Daniels, *Men Must Live Straight if They Would Shoot Straight* (Washington, D.C.: Navy Department, 1917), 16.

[58] Brandt, *No Magic Bullet*, 94.

troversial was compulsion. (Only Utah had a complete prohibition of vaccination, while five other states had modified prohibitions.) In 1809, the Commonwealth of Massachusetts enacted the nation's first compulsory vaccination law. By 1850 only five states had laws that made vaccination either compulsory or optional; an additional thirteen states had joined the ranks by 1900.[59] Opposition to compulsory vaccination varied in intensity from state to state and produced a pamphlet literature as voluminous and graphic in its titles as its British counterpart: "Horrors of Vaccination," "Compulsory Vaccination: or Blood-Poisoning by Law," and "Vaccination: A Folly and a Crime." In 1879 the opposition organized by forming the Anti-Vaccination Society of America—it was founded by an Englishman.

The legal climax to the opposition came with the U.S. Supreme Court ruling in *Jacobson v. Commonwealth of Massachusetts* (1905). The case grew out of a decision in 1902 by the Board of Health of Cambridge, Massachusetts, in response to an outbreak of smallpox, to require vaccination for all those who had not been vaccinated within the previous five years. Mr. Jacobson, claiming that he had had an adverse reaction to vaccination when a child, and that his son had had a similar reaction, refused to be revaccinated.

The Court dismissed Jacobson's contention concerning the efficacy and safety of vaccination. In effect the Court said that that decision was up to the states. The Court instead focused on two critical issues that continue to have relevance for all contemporary health promotion policies: state police power, and individual liberty. First, with regard to the police power, the Court acknowledged that it "has refrained from any attempt to define the limits of that power," but said that, "according to settled principles, the police power of a state must be held to embrace, at least, such reasonable regulations established directly by legislative enactment as will protect the public health and the public safety." Thus, the Court reaffirmed the legal justification—public health and safety—for disease prevention policies.

Second, the Court addressed the question of whether or not the vaccination law was an invasion of Jacobson's liberty, "unreasonable, arbitrary, and oppressive, and, therefore, hostile to the inherent right of every freeman to care for his own body and health in such way as to him seems best." The Court thought not. No person enjoys unrestricted liberty, even over his or her own body. "There are manifold restraints to which every person is necessarily subject for the common good." Then the Court joined the two points. "The good and welfare of the commonwealth, of which the legislature is primarily the judge, is the basis on which the po-

[59] See Stern, *Should We Be Vaccinated?*, 93–126.

lice power rests in Massachusetts." The Court upheld the right of Massachusetts to require vaccination. It did not provide anything more concrete than "the common good" and "public health and public safety" to determine where one draws the line beyond which government should not go to promulgate freedom-restricting, health-enhancing public policies. That issue continues today.[60]

THE SLIPPERY SLOPE OF STATE INTERVENTION: MATERNAL AND CHILD WELFARE

No early twentieth-century policy debate better illustrates and summarizes the major issues in the health promotion field than the Maternity and Infancy Protection Act of 1921, known as the Sheppard-Towner Act. And none provides a better introduction to yet another concern prompted by the public health revolution. This concern, expressed in a variety of metaphors involving wedges, slopes, and camels' noses, was with the cumulative consequences of public health promotion policy. Was there not the danger, asked critics, that once the state took it upon itself to make people more prudent there was no end to the mischief in which it might engage? Because of the scope of its impact, this concern was especially prominent in the debate on the Sheppard-Towner Act.

The purpose of the act was to help educate women in proper pre- and postnatal care of their infants and themselves. This was to be accomplished by creating a Maternity and Infant Hygiene Division within the Department of Labor's Children's Bureau to administer funds and provide information to the states to help guide them in carrying out health and educational programs for women. The programs included creation of maternal and infant health centers, improved birth registration procedures, dissemination of literature, and visiting-nurses programs, as well as research into issues relating to infant and maternal well-being, for example, milk supply, midwives, and maternity homes.[61] The law was innovative not only because of its purpose, but because part of the funding would come from the federal government, in the first peacetime federal grant-in-aid to the states for health care in the nation's history.

In 1912 Congress created the Children's Bureau within the U.S. Department of Labor. Among its duties was to collect data on infant and child health. The bureau issued annual reports that included infant and maternal mortality and morbidity rates, heights and weights of a national sample of children, and results from nutrition surveys. The image that

[60] See *Jacobson v. Massachusetts* 197 U.S. 11 (1905).

[61] See James A. Tobey, *The National Government and Public Health* (Baltimore: Johns Hopkins University Press, 1926), 236–43.

emerged from these reports was not pleasant: every year between 230,000 and 250,000 infants died within their first year of life, and about one-half of these deaths could be attributed to preventable causes. There was, in addition, a strong correlation between poverty and infant mortality.[62] Finally, when the U.S. infant mortality rate was compared to that of other countries, we fared poorly. One often-quoted study showed that the United States ranked last among nine comparable nations (see Table 2-3).

The maternal mortality rate was no better with the United States ranked last among twenty-one comparable nations: the U.S. rate was 8.0 per 1,000 births compared to 2.4 for the Netherlands, which ranked first. The bureau found that 80 percent of the expectant mothers in this country received no training or medical advice on matters relating to their pregnancy, and that each year between fifteen thousand and eighteen thousand women died during childbirth from preventable causes.[63]

In the 1917 bureau report, Julia Lathrop, the head of the agency, proposed a plan for maternity and infancy protection that would be administered by the bureau. Women would be educated on how to take care of themselves during pregnancy, and their infants afterward. The plan was

TABLE 2-3
Infant Mortality Rates, Selected Countries, Circa 1920 (Per 1,000 Births)

Country	*Year*	*Rate*
New Zealand	1920	51
Norway	1917	64
Australia	1920	69
Sweden	1916	70
Netherlands	1920	73
England and Wales	1920	80
Union of South Africa	1919	82
Switzerland	1919	82
United States[a]	1920	86

Source: Children's Bureau, *U.S. Department of Labor*, reported in Dorothy Brown, *The Case for the Acceptance of the Sheppard-Towner Act* (Washington, D.C.: National League of Women Voters, 1923), 5.

[a] From birth registration areas. Not all states kept birth records at this time.

[62] See J. Stanley Lemons, "The Sheppard-Towner Act: Progressivism in the 1920s," *Journal of American History* 55 (March 1969): 776.

[63] Ibid., 777.

developed into legislative form and introduced into Congress in 1918 by Jeanette Rankin, the first female member of Congress. The bill was assigned to the House Labor Committee, where it died. In 1919 it was reintroduced in the Senate by Senator Morris Sheppard (D-Texas) and in the House by Representative Horace Towner (R-Iowa). Sheppard and Towner kept the bill on the congressional legislative agenda through 1920 and 1921, during which time they held public hearings and listened to strong opposition to the bill, especially from the medical establishment. Finally, however, the bill, with the support of a wide variety of women's and other groups, was approved by large margins in the Senate in July 1921 (75 to 6), and in the House in November (279 to 39). President Harding, who had given the bill lukewarm support, signed it into law on November 21, 1921.

Proponents of the Sheppard-Towner Act began with what they believed to be the most compelling argument in favor of its support, namely, the largely preventable "appalling waste," "useless destruction," and "frightful mortality" of infants and mothers.[64] The bill's supporters reminded their colleagues of the number of infants (over 200,000) and mothers (over 20,000) who needlessly died each year. They noted that childbirth killed more women each year than any disease except tuberculosis. Furthermore, "there is hardly a country of any consequence in Europe or Asia where it is not safer to become a mother than in the United States." To prevent this waste of lives was to engage in "the noblest possible effort."[65]

The tragedy of infant and maternal mortality was decried for not only humanitarian but practical reasons as well. "This awful toll is an economic question, my friends, as well as a sentimental one." "Every child that we lose" was, according to one representative, "an economic asset to the Nation." In addition, as in the debate over venereal disease control, congressmen were reminded of the national defense implications of the bill: "Healthy mothers bear healthy children, and with healthy children we shall, in part, have done with the disgrace of inefficiency, as disclosed by our recent [World War I] draft. Should this bill assist in eventually reducing the death rate of mothers and infants 50 per cent . . . we would at once produce more healthy boys for soldiers and more healthy girls for Red Cross nurses, and at the same time more nearly approximate 100 per cent men and women for the far better pursuits of peace." In sum, Sheppard-Towner was right, both morally and pragmatically.

Major support for the bill came overwhelmingly from women's groups

[64] *Congressional Record* (28 June 1921), 3145; (19 November 1921), 7988; (28 June 1921), 3144.

[65] Ibid. (19 November 1921), 7988–99.

including, but not exclusively, remnants of the suffrage movement. Proponents and opponents agreed that fear of political reprisals from newly enfranchised women led to the overwhelming passage of the Sheppard-Towner Act. One reluctant supporter said, "Some of us remember the slaughter of Democrats in November a year ago when you Republicans slew Democrats. . . . But all previous slaughters will look like children's tea parties when the politicians really incur the righteous wrath of American womanhood." And, according to an opponent of the bill, "The woman's suffrage organizations, having accomplished their purpose in amending the Constitution, were still organized, had plenty of money, were all dressed up with nothing to do and no place to go. They clasped this baby bill to their bosoms and rushed on to Washington."[66] The *Journal of the American Medical Association* (*JAMA*) summarized the bill's passage thus: "There were just two factors: the women's vote and the politicians' fear."[67]

It is, I think, a measure of the strength of the feminist movement that five years later, in 1927, when the fear of a "women's vote" had all but disappeared, supporters could only obtain a two-year renewal of the act. Then, in 1929, it was discontinued entirely.

Opponents obviously were not against infants and mothers. Nor, for the most part, did they seem inclined to dispute the contention that a significant public health problem existed. Instead most focused their objections on the appropriateness of federal action, and the dangerous precedent this act would set. In addition, there were certain fringe ideological concerns expressed.

One objection, shared by moderates and conservatives alike, was that infant and maternal care was not an appropriate function of the federal government. According to Governor Nathan Miller of New York, "I have no quarrel with the object sought by the sponsors of the Sheppard-Towner bill." He argued, however, "that work can only be done effectively and economically by local agencies, public or private, and the will to do it, if it is to accomplish any good, must spring from local spirit and enterprise."[68] This view was put more succinctly by a U.S. congressman who articulated a fundamental facet of the American political culture: "What is done from a distance is never done well."[69] In addition to the general reservation about federal intervention was a more specific one concerning the area of public health that critics felt had traditionally belonged to states and localities. "The substantial objection to this measure

[66] Ibid. (19 November 1921), 7999–8010.

[67] *JAMA* 78 (1922): 435; Quoted in Kristine Siefert, "An Exemplar of Primary Prevention in Social Work," *Social Work in Health Care* 9 (1983): 94.

[68] *New York Times*, 22 January 1922.

[69] *Congressional Record* (23 November 1921), 8943.

is that it puts on the Federal Government a duty which under the theory of our Constitution belongs exclusively to the State."[70]

Closely related to the states' rights position was the concern that once the federal government became involved in this sort of enterprise there would be no end to it. As one representative rather colorfully stated it, "The Arab from experience knows that as soon as the camel can find a hole large enough to get his nose under the tent he will get his whole body under in time. The best place to strangle this is now."[71] Others described the bill as the "thin edge," or "entering wedge." (In the 1980s the metaphor of the camel's nose would be replaced by the "slippery slope," but the sentiment and concern would remain the same.) Some saw this "thin edge" as "socialism," "bolshevism," or "Russianization." Missouri Senator James Reed, perhaps the most vocal critic of the Sheppard-Towner Act, argued that "the fundamental doctrines on which the bill is founded were drawn chiefly from the radical, socialistic, and bolshevistic philosophy of Germany and Russia."[72] Others, like Congressman Cockran, saw the "entering wedge" as "paternalism, which is tyranny."

The threat of repeated state incursions into what Senator Reed described as "the most private and sacred relations of life" made this issue, much like the controversy over venereal disease policy, particularly sensitive.[73] As one critic warned his colleagues, Sheppard-Towner "opens the door to all kinds of vicious propaganda, including birth control and sex hygiene."[74] Opponents repeatedly warned of government bureaucrats barging into the boudoir of a pregnant or nursing mother and meddling in her most intimate behavior.

One group outside of Congress that lobbied against the bill was the state and national leadership of the medical profession. Some within the profession placed the blame for this unwarranted assertion of state power on the "endocrine perverts [and] derailed menopausics," in the Children's Bureau.[75] In the main, however, leaders of the profession expressed more dispassionate concerns. *JAMA* took a states' rights position: "The care of mother and child is a state and local, not a federal function."[76] The association also argued that the problem of infant and maternal mortality was essentially a health problem, not a sociological one, and should be left to physicians.

[70] Ibid. (19 November 1921), 8007.

[71] Ibid. (19 November 1921), 7985.

[72] Ibid. (23 November 1921), 8760.

[73] Ibid.

[74] *New York Times*, 9 March 1922.

[75] *Illinois Medical Journal* 39 (1921): 143, quoted in Siefert, "An Exemplar," 95.

[76] "Federal Care of Maternity and Infancy: The Sheppard-Towner Bill," *JAMA* 76 (5 February 1921): 383.

But the major source of concern of the profession stemmed from its own version of the "thin edge of the wedge." Physicians saw the Sheppard-Towner Act as the first step down the dark road toward "state medicine." As J. Stanley Lemons has noted, the medical profession had generally supported the Progressive movement through World War I. The two parted company, however, over the issue of national health insurance, which the American Medical Association (AMA) condemned in 1920. Thereafter, the AMA and state medical associations took the lead in opposing any actions that hinted of state involvement in health delivery. Sheppard-Towner was the first national issue over which the profession fought this battle.[77] It should be noted that the medical profession continued to oppose Sheppard-Towner and, with the decline in the perceived strength of the women's movement, was instrumental in its demise in 1929.

In order to receive the funds authorized by the Sheppard-Towner Act, each state had to formally accept provisions of the act, approve matching funds, establish a plan and authorize an agency to carry out the plan. Thus each state became a potential battleground for opponents, mainly anti-suffragists and state medical associations, to turn their national defeat into local victories. Opponents continued to charge that the act was a violation of both states' rights and individual privacy, despite the fact that the law specifically prohibited any state or federal agent from entering a home without a parent's permission. Despite these efforts the majority of the states quickly accepted the act: forty-one states passed enabling legislation in 1922. Only three states ultimately rejected participation—Connecticut, Illinois and Massachusetts, with the last unsuccessfully testing the law before the Supreme Court in 1923. In several states, including Washington (1923), Louisiana (1924), Rhode Island (1925), Vermont (1926), Maine, and Kansas (both 1927), it took a change in administration or legislative leadership before the act was accepted.

As noted earlier, the law came up for renewal in 1927 and was extended for two years. With the dwindling power of the feminist movement, and the concomitant decline in politicians' fear of the women's vote, Sheppard-Towner was not renewed, despite documentation of a decline in infant mortality from 75 to 64 per 100,000 births and in maternal mortality from 67.3 to 62.3 per 100,000 between 1922 and 1927.

The Sheppard-Towner Act was important for several reasons. First, it was the first time the federal government became involved, during peace

[77] See Lemons, "The Sheppard-Towner Act," 776–86; Rosemary C. R. Taylor, "State Intervention in Postwar Western European Health Care: The Case of Prevention in Britain and Italy," in *The State in Capitalist Europe*, ed. Stephen Bornstein, David Held, and Joel Krieger (London: George Allen and Unwin, 1984), 107–11.

time, in health promotion policy. Using the "general welfare" clause of the Constitution as its basis, Congress accepted the principle that the federal government had a role in promoting the health of its people. The law was also important because it set a precedent for federal and state collaboration in this area, and the use of federal funds as an incentive for such collaboration. The debate over the law provides students of health policy with a preview of the issues that would be raised later in the century, when the federal government would use financial and other incentives to convince states to adopt mandatory seat-belt laws and raise the minimum drinking age.

My purpose in the chapter has been to illustrate the emerging role of the state, beginning in the nineteenth century, in the area of disease prevention and health promotion through reference to some of the more prominent public health problems of the time. There were, of course, other equally pressing and controversial health issues and concerns, including tuberculosis, yellow fever, alcohol abuse, and working conditions in mining and industry. But each of these involved the same basic issue of the appropriate role of government in dealing with morbidity and premature mortality. I have tried to show that British and American governments began taking a rather expansive view of their responsibilities in these areas largely in response to the conditions attending industrialization and urbanization in each country.

It was by defining public health problems in terms of negative externalities such as the impact of ill health on national defense, economic productivity, the purity of the race, and increased public spending on the poor, or by cloaking the issues in moral garb that British and American governments could justify state involvement. This involvement was facilitated by the efforts to trace the origins of the harm to health to personal irresponsibility. I have described how the Report on Physical Deterioration attributed at least some of the blame for the poor physical condition of military recruits on the negligence of the working class itself, just as cholera and venereal diseases were viewed as manifestations of God's disfavor with personal, moral failings. (It was certainly easier to justify state intervention in those circumstances where harm is self-inflicted *and* produces negative externalities.) Finally, I have suggested that the absence of curative medical alternatives to most of the major health problems of the period further facilitated state involvement in health promotion.

With the themes of personal morality and intimacy, individual freedom, and the appropriate role of the state, along with concerns about externalities, etiology, and policy alternatives, it is possible to establish the lineage of current health promotion or disease prevention conflicts concerning AIDS, drug testing, highway safety, and restrictions on smoking, to name a few, from these earlier issues. The overriding political issue of where one draws the line between public and private responsibility for health remains as problematical today as it did in the nineteenth century and the early part of this century.

Clearly the boundary problem was less severe in the United States in the beginning of the twentieth century, by which time more substantial progress had been made in both medical science and the professionalization of the medical community than in earlier years.[78] Nevertheless, for the most part disease prevention and health promotion took place outside of, and in some instances in face of opposition from, the medical community. Health promotion, even when it may be state sponsored and seemingly coercive, becomes more acceptable when medicine cannot provide cures for the dominant illnesses that plague society. This point is well illustrated in the current debate over life-style modification policies. I turn now to that story.

[78] Paul Starr, *The Social Transformation of American Medicine* (New York: Basic Books, 1982).

Three

A New Perspective on Health: The Second Public Health Revolution

Post-Industrialization and the Second Public Health Revolution

THE FIRST public health revolution occurred in capitalist societies as a result of the changes wrought by the industrial revolution. The second public health revolution is a response to similarly dramatic changes. Some scholars use the term "postindustrial" to describe societies that have undergone or are undergoing these changes. Some academics disagree over whether or not recent social and economic changes constitute a distinct phase in capitalist development, but this particular debate need not detain us.[1] For the purpose of this book it is simply necessary to acknowledge the structural changes in British and American society since World War II, changes that have affected the nature of illness, public and private attitudes and perceptions of disease, and health promotion.

However one characterizes these changes, it is important to recognize that they have not affected both nations evenly. Most who subscribe to the notion of postindustrialization as a distinct stage in capitalist development place the United States clearly in this category, but it is somewhat less certain that Britain has reached all the requisite thresholds in education, labor force distribution, or residential living patterns to qualify for membership in this club. It is certainly the case that a number of social changes, including mass ownership of automobiles and television sets, an active feminist movement, increased proportion of the population achieving higher education, and over one-half of the workforce in nonindustrial jobs either occurred later in Britain than in the United States or were not as extensive. The magnitude and timing of the changes may differ, but their impact on both societies has been substantial. Arthur Marwick calls the period of the mid-1950s to the early 1970s "the new age of cul-

[1] See Daniel Bell, *The Coming of Post-Industrial Society* (New York: Basic Books, 1973); Christopher Lasch, "Toward a Theory of Post-Industrial Society," in *Politics in the Post Welfare State: Responses to the New Individualism*, ed. M. Donald Hancock and Gideon Sjoberg (New York: Columbia University Press, 1972), 36; Peter N. Stearns, "Is there a Post Industrial Society?" *Society* 11 (May 1974): 10–25; and Daniel Bell, "Reply to Peter N. Stearns," *Society* 11 (May 1974): 11, 23–25.

tural change" in Britain, and views this period as profoundly different from previous eras.[2]

Life and Death in Postindustrial Societies

CHANGING PATTERNS OF DISEASE

Around the turn of the century the leading causes of death and illness in the United States and Great Britain were infectious diseases such as pneumonia, influenza, tuberculosis, and gastrointestinal diseases (for example, diarrhea and enteritis) (see Tables 3-1 and 3-2). These diseases were contained during the first public health revolution through improved public sanitation and regulation of the nations' food and water supplies. Later on these infectious diseases, along with others such as diphtheria, enteric fever (typhoid and para-typhoid), and poliomyelitis, were either eradicated or their complications successfully controlled through the introduction of vaccines, sulphonamides, antibiotics, and other therapeutic inter-

TABLE 3-1
Leading Causes of Death, Ranked, United States, 1900 and 1988

Rank	*1900* *Cause of Death*	*1988* *Cause of Death*
1	Influenza and pneumonia	Heart disease
2	Tuberculosis	Malignant neoplasms
3	Gastritis	Cerebrovascular disease
4	Heart disease	All accidents
5	Vascular lesions affecting central nervous system	Pulmonary disease
6	Chronic nephritis	Pneumonia and influenza
7	All accidents	Diabetes mellitus
8	Malignant neoplasms	Suicide
9	Infant diseases	Cirrhosis and liver disease
10	Diphtheria	Atherosclerosis

Source: Monroe Lerner, "Mortality and Morbidity in the United States as Basic Indices of Health Needs," in "Meeting Health Needs by Social Action," *Annals of the American Academy of Political and Social Science* 337 (1961): 4; National Center for Health Statistics, *Monthly Vital Statistics Report* 37, no. 1 (April 25, 1988): 8-9.

[2] Arthur Marwick, *British Society Since 1945* (London: Allen Lane, 1982).

TABLE 3-2
Leading Causes of Death, Ranked, Great Britain, 1900 and 1986

Rank	*1900* *Cause of Death*	*1986* *Cause of Death*
1	Influenza and pneumonia	Heart disease
2	Bronchitis	Neoplasms
3	Lung disease	Cerebrovascular disease
4	Heart disease	Pneumonia and influenza
5	Disease of old age	Pulmonary disease
6	Cancer	All accidents
7	Diarrhea and dysentery	Diabetes
8	Infant diseases	Nephritis
9	Apoplexy	Suicide
10	Convulsions	Cirrhosis and liver disease

Sources: *Sixty-Third Annual Report of the Registrar-General of Births, Deaths, and Marriages in England (1900)* (London: His Majesty's Stationery Office, 1900), ixvii–lxxvi; *Forty-Seventh Detailed Annual Report of the Registrar-General of Births, Deaths, and Marriages in Scotland [Abstracts of 1901]* (London: His Majesty's Stationery Office, 1901), 48–49; *Thirty-Seventh Detailed Annual Report of the Registrar-General (Ireland) Containing a General Abstract of the Numbers of Marriages, Births, and Deaths Registered in Ireland during the year 1900* (London: His Majesty's Stationery Office, 1900), 2. United Nations Department of International Economic and Social Affairs, Statistical Office, *United Nations Demographic Yearbook 1986, Thirty-Eighth Issue Special Topic: Natality Statistics* (New York: Publishing Division, United Nations, 1986), 417.

ventions. There are few people in either country today who die, or suffer severe debilitation from, the infectious diseases that plagued people one hundred years ago.

The three major causes of death today in both countries are heart disease, cancer, and stroke. In terms of their health (and political) implications, these diseases play a role parallel in importance to that of infectious diseases at the turn of the century. In the United States the current age-adjusted death rate from heart disease, our number-one killer, is about 183 per 100,000 people, compared to the 1900 death rate of 202 per 100,000 for influenza and pneumonia, the major cause of death then. Similarly, cancer takes about as many lives today (about 132 per 100,000) as diarrheal diseases did in 1900 (about 143 per 100,000).

The change in the mortality and morbidity profiles of people in postindustrial countries is partly due to the success of the first public health revolution and the subsequent achievements in medical intervention in the battle against infectious diseases. This success, especially in the areas of infant and maternal mortality rates, has resulted in longer life expectancy in both countries. In the United States life expectancy increased from 47.3 years in 1900 to 75 years in 1987; in Britain the increase over the same time period has been from just under 50 years to just over 75 years. In 1900 there were three million Americans, or four percent of the population, 65 years of age or over; in 1986 there were over 29 million people, or 12.1 percent of the population, in this age group.[3] In the United Kingdom the number of people age 65 years and older increased from nearly 2 million (4.8 percent of the total population) in 1900 to 8.8 million (or just over 15 percent) in 1987.[4] The longer people live, the more likely they are to fall prey to chronic and degenerative diseases such as cancer or heart and cerebrovascular diseases, and to suffer from particular types of accidents—for example, older people are more likely to be involved in automobile or pedestrian accidents than other age groups.

CHANGING SOCIAL PATTERNS

In addition to living longer lives, most people today are living different kinds of lives from their grandparents or great-grandparents. For the most part they are better educated and more affluent; work fewer hours per week, and are more likely to spend those hours in white-collar and service positions rather than blue-collar jobs; consume more calories, which they get from different types of foods; are more likely to own their own home; and live lives in which the social and economic distance between themselves and those up and down the social ladder is less significant. Many of these changes have had important health consequences. For example, greater affluence led to dramatic increases in private automobile ownership, which increased in Britain from just under 2 million in 1938 to 6.3 million in 1961 and 17.8 million in 1987, while in the United States the increase over about the same period was from 25.3 million to 139 million passenger cars.[5] Not surprisingly, automobile deaths and injuries have become a major contemporary health problem in both nations.

[3] United States Bureau of the Census, *Historical Statistics of the United States, Colonial Times to 1970*, part 1 (Washington, D.C.: Government Printing Office, 1975), 55; and *Statistical Abstract of the United States: 1989* (Washington, D.C.: Government Printing Office, 1989), 36.

[4] Central Statistical Office, *Social Trends 18* (London: Her Majesty's Stationary Office, 1988), 24, and; *Social Trends 10*, 64.

[5] Central Statistical Office, *Social Trends 18* (London: Her Majesty's Stationery Office, 1988), 147; and Bureau of the Census, *Statistical Abstract of the United States 1989, 109th edition* (Washington, D.C.: Government Printing Office, 1989), 596.

Increased affluence also has had an impact on diet and leisure-time activity. Around 1910 Americans consumed about 120 grams of fat and 490 grams of carbohydrates per day. In 1985 fat intake was just under 200 grams and carbohydrates down to about 410 grams. Since the 1940s there has been a decline in the per capita consumption of fresh fruits, vegetables (especially potatoes), and wheat flour, and an increase in consumption of refined sugar, meats, fats, and oils. Dietary fat accounts for about 37 percent of the caloric intake of Americans, although this trend has recently been reversed.[6] On another measure of social change, the proportion of families with two wage earners has increased dramatically and has resulted in far less time spent on preparing meals and far greater consumption of precooked or convenience foods. Greater affluence and increased leisure time generally also has meant more sedentary lives and in some instances more hazardous leisure-time activities (for example, water skiing or hang-gliding).

Changes in life-styles and life chances have been experienced by virtually all groups in society, but some have been more dramatically affected than others. Among those most significantly affected from a health perspective have been women. The development and widespread use of contraceptive devices has had profound implications for the social, sexual, and economic role and behavior of women. Women in both countries are marrying later, are more likely to remain single and to have children outside of marriage, and to spend a smaller proportion of their total years of life raising children. A particularly important change has been the increased presence of women in the workforce. Between 1940 and 1986 the proportion of the U.S. workforce consisting of women increased from about one-third (33.9 percent) to over one-half (55.3 percent). In 1940 about 27 percent of all women were employed; by 1987 over 55 percent were. In Britain in 1951 about 22 percent of married women worked outside the home; by 1987 the proportion had increased to 50 percent. In addition, women are more likely than men to have a second job.

Not only are more women in the workforce, but the demographic profile of women currently working also has changed. In 1940, 64 percent of all employed women in the United States were single, widowed, or divorced. By 1987 the situation had been reversed; nearly 60 percent of the working women were married. Furthermore, increased employment rates have been greatest among women with children. In 1948 only about 11 percent of all married women with children under six years of age were in the labor force; by 1987 nearly 60 percent of all married working

[6] See United States Bureau of the Census, *Social Indicators 1976* (Washington, D.C.: Government Printing Office, 1977), 208.

women had children. In Britain about one-third of working women had a child under five years old in 1987.[7]

The changes in the role of women in British and American society are relevant to the nature of disease patterns and public health policy for two reasons. First, with the entry of more women into the labor force, and their accompanying economic power and independence, producers of goods such as alcohol, tobacco, and automobiles have begun to specifically target women. Second, many women today must cope with both the traditional demands of homemaking and child-rearing and the more recent ones of jobs and careers. This in turn has produced extraordinary tensions that have been partly assuaged by the adoption of hazardous habits including smoking and alcohol and drug use. The health consequences have been dramatic. Lung cancer among women, which was virtually nonexistent in 1940, recently passed breast cancer as the single largest cause of premature female death. Trends in alcohol- and drug-related problems parallel those for smoking and also reflect the consequences of changing life-styles among women. One indication of the impact of gender role change is that British women between the ages of 18 and 59 years old who work outside the home are more likely to be moderate to heavy drinkers (39 percent) than are homemakers (24 percent).[8] As more women move into the labor market, alcohol-related health problems have increased. For example, between 1979 and 1987 the number of British women dying from cirrhosis increased by 37 percent compared to only 11 percent among men.[9] Indeed, on a whole spectrum of life-style-related health problems, the epidemiological profile of women is beginning to look more and more like that of men.

The changes in sexual roles and practices during the post-World War II era were part of a much broader technological, social, and cultural transformation of society. Accompanying this transformation were concomitant attitudinal and value changes. Samuel Huntington characterizes these as "a new 'postbourgeois' value structure concerned with the quality of life and humanistic values, in contrast to a 'Protestant' inner-directed work ethic."[10] Similarly, Daniel Bell, the scholar most closely associated with the concept of postindustrialization, sees the role of health as central to this new value structure. He talks about an emerging "new consciousness" in which the desire to secure good health becomes a dom-

[7] See Howard M. Leichter and Harrell R. Rodgers, Jr., *American Public Policy in a Comparative Perspective* (New York: McGraw-Hill, 1984), 106.

[8] B. D. Cox et al., *The Health and Lifestyle Survey* (London: Health Promotion Research Trust, 1987), 114.

[9] *Economist*, 11 March 1989, 52.

[10] Samuel P. Huntington, "Postindustrial Politics: How Benign Will It Be?" *Comparative Politics* 6 (January 1974): 164.

inant value in postindustrial society. "The claims to the good life which the society has promised become centered on the two areas that are fundamental to that life—health and education."[11] Whatever else one may wish to accept or reject about the concept of postindustrialization, no observer of the contemporary scene would quarrel with the characterization of ours as a time in which maintaining good health has become a major personal, social, and political concern. As I suggested in Chapter 1, this concern appears stronger in the United States, especially among middle- and upper-class men and women, than in Britain. It is unclear to me whether this represents a fundamental cultural difference between the two societies, or merely a function of the more advanced stage of postindustrialization and greater affluence in the United States.

COMPARING THE TWO PUBLIC HEALTH REVOLUTIONS

Thus far I have emphasized that as people in the United States and Great Britain approach the end of the twentieth century they face a different array of social and economic choices, and health hazards, than their nineteenth century counterparts. Yet despite the differences there are important political and medical similarities between the two periods. First, although the origin and nature of the predominant diseases differ, there was a sense in both periods that these health problems were largely individually or socially inflicted, and avoidable. At the turn of the nineteenth century, the conventional wisdom was that better sewerage and refuse disposal, cleaner streets, safer milk and water supplies, not spitting in public, better pre- and postnatal nutritional care, and regulation or elimination of prostitution would substantially reduce the risks of disease. Today we are advised to eat, drink, and drive more sensibly, not smoke, exercise, and be more "responsible" in our sexual behavior. In both periods there was the sense that personal or collective irresponsibility was the key, and controllable, factor in disease prevention.

A second parallel between the two periods is the current belief that medicine and science are significantly limited in their ability to deal with the major causes of illness and early death. The major health problems were, and are, essentially public health rather than medical problems. Rudolf Klein has noted that today, as in the nineteenth century, emphasis is placed on what he calls "social engineering" rather than the "medical engineering" that dominated the post-World War I era.[12] Starting with Thomas McKeown in England, many public health officials, policymak-

[11] Bell, *The Coming*, 128.

[12] Rudolph Klein, *The Politics of the National Health Service* (London: Longman, 1984), 174.

ers, and the general public, on both sides of the Atlantic, have concluded that "modern medicine has been able to make very little headway with" cancer, heart disease, stroke, and AIDS, not to mention automobile accidents and diseases related to alcohol and drug abuse.[13] Some have questioned the "end of medicine" or "limits of medicine" thesis, but there is no question that "the progressive disillusionment with curative medicine" has played a significant role in the current emphasis on disease prevention through life-style change.[14]

Finally, the two periods are similar in that during each the health of the nation became a national issue in large part for political reasons, rather than exclusively health reasons. In the mid to latter part of the nineteenth century the state became involved in public health largely because it appeared that the national interest required that it do so. Today, too, while it is asserted that state involvement in personal health-related choices is in the interest of the individual, the emphasis in policy debates is on the collective, and particularly economic, implications of personal behavior.

The current emphasis on the socioeconomic and behavioral origins of ill health has become the focal point of health policy debates in both Britain and the United States. The second public health revolution, or new perspective, raises anew many of the issues that emerged during the first public health revolution. In the remainder of this chapter, I will examine the major issues, assumptions, and implications of the debate over health promotion and disease prevention through life-style modification.

The New Perspective on Health

The Assumptions

THE ENEMY IS US: THE ETIOLOGY OF DISEASE

The most basic assumption of the new perspective is that a good deal of disease is self-inflicted, a product of our daily habits, and that individuals,

[13] Rosemary C. R. Taylor, "State Intervention in Postwar Western European Health Care: The Case of Prevention in Britain and Italy," in *The State in Capitalist Europe*, ed. Stephen Bornstein, David Held, and Joel Krieger (London: George Allen and Unwin, 1984), 93; Thomas McKeown, *The Role of Medicine: Dream, Mirage, or Nemesis* (Princeton: Princeton University Press, 1979); Robert Evans, "A Retrospective on the 'New Perspective,' " *Journal of Health Politics, Policy and Law* 7, no. 2 (Summer 1982): 326; Victor Fuchs, "The Economics of Health in a Post-Industrial Society," *The Public Interest* (1979): 3–20.

[14] Ernest Saward and Andrew Sorenson, "The Current Emphasis on Preventive Medicine," in *Health Care: Regulation, Economics, Ethics, Practice*, ed. Philip H. Abelson (Washington, D.C.: Association for the Advancement of Science, 1978), 49. For a dissenting view on the "limits of medicine" thesis see Walsh McDermott, "Medicine: The Public Good and One's Own," *Perspectives in Biology and Medicine* 21, no. 2 (Winter 1978): 167–87.

through negligence, self-indulgence, and irresponsibility, contribute significantly to their own ill health or premature death. This assumption enjoys widespread currency among academics, physicians, policymakers, and the general public. Consider just two of the many assertions of this position:

> Much ill health in Britain today arises from over-indulgence and unwise behavior. Not surprisingly, the greatest potential and perhaps the greatest problem for preventive medicine now lies in changing behavior and attitudes to health.[15]

> Personal habits play critical roles in the development of many serious diseases and in injuries from violence and automobile accidents. Many of today's most pressing health problems are related to excesses—smoking, drinking, faulty nutrition, overuse of medications, fast driving, and relentless pressure to achieve.[16]

The premise that individuals, through personal imprudence and irresponsibility, contribute significantly to their own morbidity and premature mortality appears unassailable in view of the evidence linking various personal habits and the major causes of disease and early death. The list of diseases and contributing behavioral choices varies only marginally from one government report or professional paper to another. A synthesis of the literature produces the gruesome litany outlined in Table 3-3. Certainly the scientific evidence relating certain habits to specific diseases is stronger in some cases than in others. The precise nature of the relationship between cervical cancer and early onset of coitus or multiple sexual partners is neither completely understood nor completely accepted by the medical profession. On the other hand, only a few still doubt the adverse effects of cigarette smoking and alcohol and drug abuse.

Some continue to question the causal relationship between specific personal choices and a particular disease, but few in the general population, or among health professionals, doubt the premise "that our way of life is a major key to our sickness and health."[17]

Two major corollaries follow from this first premise: "Each of us can do more for our own health than any doctor, any hospital, any machine, or any drug;"[18] and individuals *must* be more responsible in, or made to be more responsible for, their behavior. If one accepts the first premise, namely that many of today's health problems are self-inflicted, then the

[15] United Kingdom Department of Health and Social Security, Department of Education and Welfare, Scottish Office, Welsh Office, *Prevention and Health* (1977).

[16] United States Department of Health, Education, and Welfare, *Healthy People: The Surgeon General's Report on Health Promotion and Disease Prevention* (Washington, D.C.: Government Printing Office, 1979).

[17] Leon R. Kass, "Regarding the End of Medicine and the Pursuit of Health," *Public Interest* 40 (Summer 1975): 31.

[18] Joseph A. Califano, Jr., *America's Health Care Revolution* (New York: Random House, 1986), 188.

TABLE 3-3
Life-Style And Self-Inflicted Diseases

Life-Style	*Self-Inflicted Diseases*
Alcohol abuse	Cirrhosis of the liver Encephalopathy and malnutrition Motor-vehicle accidents Obesity
Cigarette smoking	Chronic bronchitis Emphysema Lung, bladder, and breast cancer Coronary artery disease
Abuse of pharmaceuticals	Drug dependence Adverse drug reactions
Addiction to psychotropic drugs	Suicide Homicide Malnutrition Accidents
Overeating	Obesity
High fat intake	Arteriosclerosis Coronary artery disease Diabetes
Low-fiber diet	Colorectal cancer
High carbohydrate intake	Dental caries
Lack of exercise	Coronary artery disease
Failure to wear seat belts	Higher incidence of severe automobile injuries and deaths
Sexual promiscuity or "irresponsibility"	Syphilis Gonorrhea Cervical cancer AIDS
Sun tanning	Skin cancer

Sources: Adapted from Marc Lalonde, *A New Perspective on the Health of Canadians* (Ottawa: Department of National Health and Welfare, 1974). See also R. Lewy, *Preventive Primary Medicine* (Boston: Little Brown and Co., 1980); J. H. Knowles, "The Responsibility of the Individual," in *Doing Better and Feeling Worse: Health in the United States*, ed. J. H. Knowles (New York: Norton, 1977).

question arises as to what role, if any, the state should play in refashioning reckless behavior. Before addressing this question it is necessary to examine the second major premise of the new perspective.

WE CAN NO LONGER AFFORD THE COSTS OF FOOLISHNESS

The emergence of a health philosophy emphasizing greater individual responsibility in disease prevention was precipitated by two health-related developments. The first, which has been discussed already, was the increased recognition that there are severe limits to the ability of medicine to conquer many of our most serious health problems. The second is the explosion in health care costs. In the case of the United States, health care costs have increased from a total of $12.7 billion a year, representing $81.86 per person and 4.4 percent of the gross national product in 1950, to an estimated $590 billion or $2,306 per person in 1989, over 11 percent of the gross national product. Furthermore, the annual increases in health costs have far exceeded those of any other good or service in the Consumer Price Index; overall inflation was between 4 and 5 percent during 1986–1987, while health care costs increased 9.8 percent. Increases in Britain, on a much smaller scale, have paralleled those in the United States. Health care costs have increased from 3.9 percent of the gross national product in 1950 to 6.2 percent in 1983, a per capita increase of about £40 to £120 ($95.26 to $285.70).

It would be erroneous, of course, to conclude that high and rapidly increasing national health care costs are solely the result of harmful lifestyles. Other factors, including an aging population, overall inflation, costly advances in medical technology, the structure of health care financing mechanisms, and, in the United States, over-doctoring and defensive medical practices in response to a malpractice crisis, have all contributed to the explosion in health care costs.[19] Nevertheless, many policymakers would agree that "the greatest portion of our national [health] expenditure goes for caring for the major causes of premature, and therefore preventable, death and/or disability. . . ."[20] And, although no specific price tag has been placed on the costs of premature or preventable mortality and morbidity due to imprudent habits, they run into the tens and perhaps hundreds of billions of dollars. In the United States the direct health care costs associated with smoking and alcoholism have been estimated to be in excess of $20 billion per year.

The influence of soaring health care costs, and the impact of life-style factors on these costs, was identified in an early British government report

[19] See Leichter and Rodgers, *American Public Policy*, 82–88.

[20] John H. Knowles, "The Responsibility of the Individual," in *Doing Better and Feeling Worse: Health in the United States*, ed. John H. Knowles (New York: Norton, 1977), 75.

on health promotion. In 1977 a committee of the House of Commons examined the "problem" of increasing health care costs. The resulting report began by noting that "this enquiry was undertaken in response to concern at the high and increasing cost of the National Health Service." The report went on to acknowledge that "the Sub-Committee was particularly interested in the prevalence of disease precipitated by the individual's habits." It concluded that "we have been convinced by our enquiry that substantial human and financial resources would be saved if greater emphasis were placed on prevention."[21] The issue of rapidly rising health care costs is particularly critical in Britain because over 95 percent of all money spent on health care comes from the government, compared to 40 percent in the United States. What this means is that in both countries, but more so in Britain, health care competes directly with other public services for scarce resources.

British policymakers have extolled the virtues of greater individual responsibility in health not only because it allegedly will save money, but because it will allow the remaining resources to be used more deservedly. According to the Conservative government's policy handbook *Care in Action*, "A general aim should be to help people appreciate that much illness is avoidable and that avoidable illness preempts resources needed for the treatment of those who are unavoidably sick."[22] There is an almost Victorian quality to the implicit admonition that not to take better care of oneself is selfish.

The notion that disease prevention is the answer to the health care cost crisis is shared by American policymakers as well. President Jimmy Carter, in his introduction to *Healthy People*, said that disease prevention "can substantially reduce both the suffering of our people and the burden on our expensive system of medical care."[23]

It is not at all surprising, then, that financially strained nationalized health delivery systems, as in Britain, and private systems with a large subsidized component, as in the United States, would enthusiastically embrace a health promotion approach to health policy.

Criticisms of the New Perspective

Despite near-universal agreement that certain life-style changes would help reduce the human and material costs of illness and premature death,

[21] United Kingdom Expenditure Committee, *Preventive Medicine* (London: Her Majesty's Stationery Office, 17 February 1977).

[22] United Kingdom Department of Health and Social Security, *Care in Action* (London: Her Majesty's Stationery Office, 1981), 12.

[23] United States Department of Health, Education, and Welfare, *Healthy People*, introduction.

many people have serious reservations about the new perspective. The following discussion is a general one. In subsequent chapters I will show how these general concerns manifest themselves in specific policy debates.

DILEMMAS OF UNCERTAINTY

I have already suggested one objection to the new perspective, namely that the evidence upon which life-style-related policy proposals are based is often inconclusive or ambiguous. According to a British government document, "The recurring dilemma is whether or not to act on the basis of inadequate information."[24] This dilemma derives from three factors. First, evidence connecting a particular health problem and life-style pattern is often hard to acquire, partly as a result of the practical and ethical difficulties involved in conducting clinical research on human subjects. One witness before the Parliamentary committee on "Preventive Medicine" noted with regard to the relationship between smoking and heart disease that "one of the problems in this field . . . is that it is not possible to do the kinds of experiments in terms of proving the value and effectiveness of a particular measure that people have become accustomed to."[25]

The dilemma facing both citizens and policymakers is exacerbated when the scientific community is itself either uncertain or divided about the causes of particular health problems. The ambiguity of the scientific message must surely have frustrated the members of the parliamentary committee on "Preventive Medicine," who were expected to make policy recommendations based upon contradictory evidence, one scientist arguing that sucrose was the main dietary cause of heart disease, even as another attributed the problem to dietary fat intake.[26] This problem is especially evident in areas dealing with personal habits and various forms of cancer. Since there are no definitive explanations of carcinogenesis, challenges to assertions of causality abound. Most relevant to this book is, of course, the issue of smoking and ill health, but perhaps the most fertile field for evidential squabbling involves environmental and workplace-related diseases.[27]

[24] United Kingdom Department of Health and Social Security, *Prevention and Health: Everybody's Business* (London: Her Majesty's Stationery Office, 1976), 66.

[25] *Preventive Medicine*, xx.

[26] *Preventive Medicine*, xxi.

[27] See Troyen A. Brennan and Robert L. Carter, "Legal and Scientific Probability of Causation of Cancer and Other Environmental Disease in Individuals," *Journal of Health Politics, Policy and Law* 10 (Spring 1985): 33–80. For a particularly critical position on the causal analysis of the smoking and lung cancer debate see David Collingridge and Colin Reeve, *Science Speaks to Power: The Role of Experts in Policy Making* (New York: St. Martin's, 1986), especially 123–44.

A third, and related, part of the dilemma is that neither citizens nor policymakers are typically equipped to sort out conflicting scientific evidence and make a rational personal or policy decision. The conflict between science and policy-making, to which I refer in Chapter 1, varies among and within policy areas. It poses a greater problem in health or environment policy than other areas of public policy since the former relies more heavily upon information that is often inaccessible and unintelligible to the lay person. In addition, most Americans will feel more competent to judge the evidence concerning the relationship between seat-belt use and the likelihood of injury than between anal sex and AIDS.[28]

Another dimension of this etiological dilemma is how one determines whether a particular health problem is self-inflicted or the result of involuntary, capricious factors. This question is critical to the policy-making process because it helps color the nature of the policy response. In the case of both alcoholism and AIDS, for example, there is considerable public debate over whether these are the result of reckless behavior or random biological chance. Clearly society, and public policymakers, respond differently to innocent victims of disease than they do to those who have caused their own ill health through foolish (or perhaps immoral) life-style choices.

The dilemma facing policymakers in this environment of uncertainty is obvious: "Clearly," according to the authors of the British white paper *Prevention and Health*, "the Government cannot wait for absolute certainty before taking action; on the other hand, it would seem wrong to use resources or interfere with individual liberty if there is real doubt whether the action proposed will in fact produce the benefits claimed for it."[29]

THE THREAT TO INDIVIDUAL LIBERTY

As the British government white paper mentioned above suggests, the need to proceed cautiously because of uncertain scientific evidence is especially compelling since individual liberty is often at stake. This concern has dominated much of the debate over life-style legislation and has been

[28] For a discussion of the problems associated with the role of science in policy-making see Duncan MacRae, Jr., "Science and the Formation of Policy in a Democracy," *Minerva* 11 (1973), 228–42; see also Allan Mazur, "Dispute Between Experts," *Minerva* 11 (1973), 243–62; and Jack DeSario and Stuart Langton, "Citizen Participation and Technocracy," *Policy Studies Review* 3 (February 1984), 223–33. This issue of the *Policy Studies Review* contains a "Symposium On Citizen Participation and Public Policy." For a general discussion of the abandonment of political judgment by citizens to experts see Ronald Beiner, *Political Judgment* (Chicago: University of Chicago Press, 1983).

[29] United Kingdom Department of Health and Social Security, *Prevention and Health*, 6.

particularly troubling to conservatives. The problem is nicely summarized in a statement by a former member of the Thatcher government:

> How far are a government justified in taking action to seek to persuade people to abandon a course of conduct which that Government believe to be harmful? Particularly, should this question be asked of a [Conservative] Government who have been elected to reduce the interference of the State in the lives of the people and to leave individuals greater freedom to make their own decisions? I am sometimes a little envious of those outside the House who find the case for individual choice so unimportant that they would cheerfully step in and legislate, interfere, control and ban without qualm.[30]

It is in light of the freedom-limiting nature of most health promotion policies that the "free to be foolish" argument is most apparent. David Mellor, a Conservative member of Parliament, argued against mandatory seat-belt legislation by saying that "those who value a free society must willingly embrace situations where some of our fellow citizens, some of the time, are entitled to behave foolishly if they so wish."[31] Finally, conservatives are concerned that state intrusion in private affairs can become habit-forming. As one Washington State legislator expressed it in a debate on seat belts, "The minute . . . government makes that first step to infringe upon freedom, it's just a period of time before it takes more serious steps."[32]

Although liberals and social democrats are neither unmindful of nor unmoved by concern for individual freedom, they see the issues raised by life-style legislation in terms of balancing personal freedom and social need. Hence, the Labor government's statement on prevention and health noted, "The problem reflects a fundamental dilemma of democratic society. But preventive medicine has always involved some limitation of the liberty of the individual, often for his own good, often for that of the community, and often for both."[33]

Repeatedly in the course of debates over life-style modification legislation, policymakers have struggled over where to draw the line between government regulation and individual discretion. For conservatives the answer has always been that the individual should be left with as much freedom as possible to choose his or her life-style, however foolish that choice might be. While acknowledging that certain behavior patterns impose negative externalities on others, conservatives argue that—except in extraordinary circumstances such as those involving children or mentally

[30] United Kingdom, *Parliamentary Debates* (Commons), Vol. 984 (9 May 1980), col. 781.

[31] Ibid., Vol. 970, No. 43 (Road Traffic Bill) (20 July 1979), col. 2215.

[32] Washington State Senator Kent Pullen.

[33] United Kingdom Department of Health and Social Security, *Prevention and Health*, 5.

incompetent persons—the state should resort to noncoercive alternatives, like education, to affect life-style choices.

VICTIM BLAMING

Another criticism of the new perspective is the tendency of some of its advocates to place almost exclusive responsibility on the individual for ill health. Especially critical of this emphasis are American academics of the left. For example, Rob Crawford, Howard Berliner, and Sylvia Tesh have all denounced the heavy emphasis on individual responsibility for ill health as a political ploy aimed at diverting attention from both the real socioeconomic origins of disease and the failures of the health care system (for example, runaway medical costs and inequitable and inaccessible health care). Most critics on the left identify the dominant economic interests of the nation as the main culprits in this "victim blaming" or "sickness as sin" diversion. According to Sylvia Tesh, "When we place the environmental cause of disease at the far end of the [causal] chain, we condone the very limited disease-prevention practices advocated by industrialists. We buy into the idea that protecting industry takes precedence over protecting health. We opt for the disease-prevention programs which least interfere with industrial production."[34]

One need not entirely accept this critique, or its ideological implications, to recognize validity in it. First, in both the United States and Britain there are economic interests that oppose various prevention policies. Dr. David Player, the former director of the British Health Education Council, particularly singled out the tobacco, alcohol, and agricultural industries, as would not surprise many students of health politics in the United States.[35]

Second, offered in its extreme form, as it often is, the individual responsibility thesis divorces the person from his or her social environment. Assertions such as those by Victor Fuchs—"the greatest potential for improving health lies in what we do and don't do for and to ourselves. The choice is ours"—and Joseph Califano—"we have met the enemy and they are us"—strongly imply that the individual lives in a socioeconomic vacuum.[36] This view assumes complete knowledge of, and agreement on, the etiology of disease. However, significant uncertainties remain about the genesis of some diseases, and many may have multifactorial origins.

[34] Sylvia Tesh, "Disease Causality and Politics," *Journal of Health Politics, Policy and Law* 6, no. 3 (Fall 1981): 387–88.

[35] Melba Wilson, "Resistance from Within," *Health and Social Service Journal* (19 May 1983): 587.

[36] Fuchs, "The Economics of Health," 151; Califano, *America's Health Care Revolution*, 188.

There are too many diseases about which we know too little to place the blame solely on irresponsible personal behavior.

Third, the assumption here about the autonomy of the individual and the voluntariness of life-style decisions is, in some instances, dubious. Amitai Etzioni, in critiquing the "health as individual responsibility" thesis, notes that "it tends to overlook or misconstrue the nature of the societal constraints on the individual will . . . and the role social conditions have in maintaining unhealthy behavior habits."[37] One might add that social conditions are influential not only in maintaining unhealthy habits, but also in stimulating their initial adoption. In a society in which modern advertising and merchandising techniques, government crop subsidies, peer-group pressure, and the addictive nature of the substance all conspire to encourage cigarette smoking, how voluntary is this high-risk habit?

The question of autonomy and responsibility in selecting health-enhancing and eschewing self-harming behavior patterns is particularly relevant, given the evidence over the centuries relating social class to morbidity and premature mortality. At least since the days of Chadwick and Griscom it has been consistently shown that the lower someone's socioeconomic status, the greater the likelihood of disease and early death, and it is among the poorer classes in which the least autonomy and decision-making latitude exists.

The most recent detailed analysis of the relationship between social status and health is the 1980 Black Report, mentioned in Chapter 2, on inequalities in health in Britain. The major conclusion of the report is that despite thirty years of commitment and effort under the National Health Service to achieve equality of health care, inequalities in health persist among occupational, racial, and geographical groups in Britain. This conclusion was "clearest and most unequivocal" with regard to the relationship between occupational class and mortality. For instance, those persons from families of unskilled workers were from one and one-half to three times more likely to die prematurely, at every stage of life, than those from professional families. The report concluded that the most convincing explanation for these differences involved variations in the "material conditions of life" that are an integral part of the British class structure.[38] Furthermore, the report notes "the many ways in which people's behavior is constrained by structural and environmental factors *over which they have no control*."[39] Although some have criticized the types

[37] Amitai Etzioni, "Individual Will and Social Conditions: Toward An Effective Health Maintenance Policy," *The Annals of the American Academy of Political and Social Science* 437 (May 1978): 65. Emphasis added.

[38] Peter Townsend and Nick Davidson, eds., *Inequalities in Health: The Black Report* (Harmondsworth: Penguin, 1981), 207.

[39] Ibid., 212. Emphasis added.

of measurement and analysis in the Black Report, for my purpose the main point that needs to be stressed is that in its conclusions and emphasis on the social, as opposed to individual, origins of ill health, the Black Report is a direct descendant of Chadwick's *Report on the Sanitary Condition of the Labouring Population of Great Britain*.[40]

In sum, then, although the reduction of stress, increased recreational activity, and wiser eating habits all may be indicated for healthier lives, such choices may be made impossible, or at least considerably more difficult, given the educational, economic, and cultural constraints of those living at or near the poverty level. All this is not to suggest that individuals, whatever their socioeconomic group, have no responsibility for or control over unwise health-related decisions. It is to suggest the frequency and range of life-style choices available to individuals varies greatly.

The "sickness as sin" emphasis in the second public health revolution bares a striking resemblance to the nineteenth-century conception of sickness as divine judgment for immoral behavior. A secular explanation for the victim-blaming bias of the new perspective is offered by J. P. Allegrante and R. P. Sloan. Relying on M. J. Lerner's "Just World Hypothesis" derived from social psychology, they explain that "we tend to perceive the world as a just place in which people get what they deserve and deserve what they get. This applies not only to those people who are the beneficiaries of positive events, but also those who are victimized by misfortune." Hence we view those with illness as getting their just desserts because of their flawed behavior. "In this way, at least psychologically, we are protected against the possibility that we will suffer the same illness."[41] The current policy debate on AIDS is comfortably accommodated into this world view.

Although it is indisputable that certain personal habits play a vital role in the onset of illness and may result in premature death, it is important to recognize, as various critics have, that there may be other intervening factors involved. In addition, it is also important to understand that any explanation of illness, including that which places individual responsibility at its center, may serve economic and political as well as scientific interests.

IS PREVENTION WORTH IT?

One of the cardinal tenets of the current public health movement is the old proverb that "an ounce of prevention is worth a pound of cure." As President Carter said in his introduction to *Healthy People*, "Our fasci-

[40] See Roy Carr-Hill, "The Inequalities in Health Debate: A Critical Review of the Issues," *Journal of Social Policy* 16, no. 4 (October 1987): 509–42.

[41] J. P. Allegrante and R. P. Sloan, "Ethical Dilemmas in Workplace Health Promotion," *Preventive Medicine* 15 (1986): 314–15.

nation with the more glamorous 'pound of cure' has tended to dazzle us into ignoring the often more effective 'ounce of prevention.' "[42]

Recently this assumption has been challenged. In a book entitled *Is Prevention Better Than Cure?* Louise Russell of the Brookings Institution examined a variety of disease prevention procedures or approaches, including smallpox and measles vaccination, screening and drug therapy for hypertension, and exercise. Her major conclusion was that "even after allowing for savings in treatment, prevention usually adds to medical expenditures, contrary to the popular view that it reduces them." Wisely, Russell does not generalize from the limited but important cases that she has examined to all disease prevention measures: "Individual measures must be evaluated on their merits."[43] Others have reached a similar conclusion. Rogers, Eaton, and Bruhn, in an article called "Is Health Promotion Cost Effective?" review literature published from 1969 through 1979 and conclude that "health-promotion programs have yet to demonstrate their effectiveness or efficiency conclusively and require the dual evaluation of cost-effectiveness analysis."[44] Similarly, K. Michael Peddecord, in assessing the economics of risk reduction, makes the following observation:

> Although there is a widely recognized potential for cost savings from prevention, the nature of savings is unclear. Before selling prevention as a panacea for health care cost problems, cautions concerning the macroeconomics of prevention must be considered. While there exists considerable potential for improved health status and longevity gains through prevention, it is not clear that prevention of today's major causes of morbidity and mortality will reduce the costs of disease care in the long run. It must be recognized that a surge of other competing morbid conditions is likely as the population becomes older.[45]

I am not suggesting that policymakers should discount the importance of prevention, or reject the utility of preventive health policies. Nevertheless we should be alert to the tendency of the new perspective advocates to place cost-effectiveness at the center of health promotion proposals. Both Russell and Peddecord, however, remind us that the main purpose of prevention is, or should be, to minimize pain and suffering and extend

[42] United States Department of Health, Education, and Welfare, *Healthy People*, Introduction.

[43] Louise Russell, *Is Prevention Better than Cure?* (Washington, D.C.: The Brookings Institution, 1986), 110–11.

[44] Peggy Jean Rogers, Elizabeth K. Eaton, and John G. Bruhn, "Is Health Promotion Cost Effective?" *Preventive Medicine* 10 (1981): 338.

[45] K. Michael Peddecord, "Competing for Acute Care Dollars: The Economics of Risk Reduction," in *Promoting Health through Risk Reduction*, ed. Marilyn Faber and Adina Reinhardt (New York: Macmillan, 1982), 325.

healthful life: "For even when prevention does not save money, it can be a worthwhile investment in better health, and this—not cost saving—is the criterion on which it should be judged."[46]

Policy Responses to the New Perspective

Since the mid-1970s, both the United States and Britain have incorporated health promotion into their national health policies. The commitment has been articulated in key public reports, has received increased budgetary attention, and has been implemented in specific legislation. In the United States, health promotion has become a concern of states and localities, as well as the federal government. In Britain, the main policy initiative has come from the central government, with much of the implementation for traditional disease prevention activities such as immunization and pre- and postnatal counseling and care the responsibility of local and regional health authorities.

Since I will examine specific life-style modification policies in Chapters 4 through 7, in the remainder of this chapter I will simply outline the general evolution in health promotion policy in each country. In addition, I will suggest some of the difficulties in evaluating the nature, extent, and material commitment to these more general health promotion efforts.

GREAT BRITAIN: EVERYBODY'S BUSINESS

British and American governments have embraced health promotion and acknowledged some state responsibility in this area. This commitment has ranged from reinforcing support for the traditional areas of disease prevention, such as immunization and health education, to supporting programs that encourage physical fitness, more "responsible" sexual behavior, and more prudent diets.

For ideological reasons, both British and American governments have favored a primarily educational approach to health promotion. The essential message, especially during the Reagan and Bush and Thatcher years has been that the state's chief obligation, and most effective role, is to provide citizens with the information necessary to make wise life-style decisions. In this regard it is instructive to note that the Thatcher government has become one of Britain's major advertisers, the third largest in 1988. Much of this advertising promotes the sale of government enterprises, but a good deal promotes self-reliance and personal responsibility,

[46] Russell, *Is Prevention Better than Cure?*, 5.

especially in the area of disease prevention. Thus, there have been ambitious AIDS prevention, prenatal care, and anti-drug use campaigns.[47]

Official British concern with disease prevention, and especially the impact of life-style on health care costs, began with a special inquiry into preventive medicine in November 1975 under the auspices of the Social and Employment Sub-Committee of the Expenditure Committee of the House of Commons. The inquiry was undertaken in response to the high and increasing costs of the National Health Service. At the outset of its report the subcommittee recognized the "growing national and international interest in the subject of preventive medicine." It then addressed the specific issue of life-style and health, indicating that it "was particularly interested in the prevalence of disease precipitated by the individual's habits . . . sometimes described to us as 'disease of civilization' or 'self-induced disease.' "[48]

The subcommittee submitted its report in February 1977. It contained fifty-eight recommendations, many of which dealt with life-style-related health problems. For example, it recommended abolishing cigarette coupons, prohibiting cigarette machines on premises to which children had access, putting stronger health warnings on cigarette packages, designating nonsmoking areas in public places, and banning tobacco advertising. There were also several proposals concerning alcohol abuse. These, however, were couched in rather permissive terms. By way of illustration, the subcommittee recommended that the legal drinking age "should in no circumstances be lowered," and alcoholic beverages "should not be allowed to become a relatively cheap item in the shopping basket."[49]

Other life-style-related proposals dealt with sex education, contraception, physical fitness, and diet. The overwhelming majority of the proposals, regardless of the area of concern, emphasized the use of health education, not regulation, as the means to achieve improvements in the health of the nation. After reviewing various strategies for prevention, the parliamentary report concluded, "Education is, we believe, the preferable way."[50] It must be remembered, however, that policy in the British systems emanates, with very few exceptions (such as, private members' bills), from the prime minister and the cabinet. In other words, the government was in no way bound to follow the subcommittee's recommendations.

Health promotion was, however, very much a concern of the govern-

[47] See *New York Times*, 23 May 1989.

[48] United Kingdom Social and Employment Sub-Committee of the Expenditure Committee of the House of Commons (London: Her Majesty's Stationery Office, 17 February 1977), xvi.

[49] Ibid., lxxxiv.

[50] *Preventive Medicine*, xxii.

ment. The parliamentary study coincided with the publication of a consultative document entitled "Prevention and Health Everybody's Business," prepared by the health departments of Great Britain and Northern Ireland and intended for mass consumption. Popularly referred to as the "Red Book" for its striking red cover, "Prevention and Health" identified three broad categories of disease: those diseases and disabilities associated with old age; hazards associated with environmental factors; and "those diseases the cause of which and the solution to which can be laid at the door of man's behavior."[51] The "Red Book" offered no policy proposals. Its purpose was "not to recommend specific programmes but to start people thinking and talking about prevention." Furthermore, although acknowledging that government policy was important in this area, the document noted that "to a large extent, though, it is clear that the weight of responsibility for his own health lies on the shoulders of the individual himself. The smoking related diseases, alcoholism and other drug dependencies, obesity and its consequences, and the sexually transmitted diseases are among the preventable problems of our time and in relation to all of these the individual must choose himself."[52]

In December 1977 the Labor government formally responded to the parliamentary report in a White Paper, "Prevention and Health." The White Paper, which in effect constituted the Labor government's official policy, espoused the view that "in the past, improvements in the health of the population derived largely from advances in environmental living conditions and measures to control infectious disease. Today, the greatest scope for further progress would seem to lie in seeking to modify attitudes and behavior in relation to health."[53] The government promised "that in the future much greater emphasis would be given to prevention." Having stated this it went on to explain the political, economic, and technical limitations associated with implementing such a program. These included many of the issues discussed above, including the concern for individual liberty, "the dilemmas of uncertainty" involving evidence and causality, and the appropriate role of the state in affecting life-style change. In the end the government fully accepted twenty-four of the subcommittee's recommendations, accepted with reservations seventeen, indicated that it had sixteen under consideration, and rejected eight. Among the proposals it rejected were those that would have required Area Health Authorities to make public the proportion of their budgets devoted to preventive services, banned tobacco advertising, abolished cigarette coupons, required

[51] United Kingdom Department of Health and Social Security, *Prevention and Health: Everybody's Business*, 31.

[52] Ibid., 38.

[53] United Kingdom Department of Health and Social Security, Department of Education and Science, Scottish Office, Welsh Office, *Prevention and Health* (1977), 10.

stronger health warnings on cigarette packages, prohibited cigarette machines on premises accessible to minors, encouraged research into the long-term effects of fluoride, and reallocated state resources from high-technology medicine to preventive medicine. The government did support almost all of the health education proposals, and it encouraged physicians, employers, and parents to play a more active role in facilitating, teaching, and practicing greater health promotion through wiser lifestyles. The emphasis, then, of the Labor government during the late 1970s was on education and individual responsibility, not regulation or coercion.

This emphasis continued with only minor variations under the Conservatives, who returned to power in 1979. Their formal position was outlined in a handbook intended primarily for officials of newly created district health authorities. The form and content of the message was a familiar one: the population was getting older, medical treatments were becoming more costly, resources were getting more scarce, and efficiency was necessary. What could be done? First, individuals had to recognize that they had "clear responsibilities" in this area. Second, the role of the state was "to give information available about risk to health and to minimize such risks by developing the services and improving the conditions that produce good health." Third, local public health authorities had a special responsibility and a unique opportunity to encourage and facilitate health promotion. "Much of the work of preventing ill health, whether by counseling, immunisation or education, has to be undertaken locally. Health authorities are best placed to know the needs of the population they serve."[54] The government then spelled out "a local strategy of health promotion and preventive medicine," which would

1. Put pressure on owners of cinemas and other public places to accept non-smoking as the norm.
2. Improve the availability of genetic counseling and family planning.
3. Improve the liaison with education authorities to monitor student health.
4. Encourage immunization.
5. Coordinate efforts of local public and private entities to encourage better diet and exercise and reduce smoking.
6. Support health education efforts in the schools.
7. Help create a favorable climate of opinion of fluoridation of water supplies.
8. Create "awareness by voluntary, community and commercial organisations of the need to harmonise their efforts to ensure that the community has a positive approach to health promotion and preventive medicine."[55]

[54] *Care in Action*, 11.
[55] Ibid., 12–13.

For nearly two decades, then, Labor and Conservative governments have extolled the virtues of disease prevention and health promotion. Despite the apparent rhetorical emphasis on the importance of prevention, it is difficult to assess the actual commitment to the task. Since no level of government has a specific budget category for "prevention" or "health promotion," it is virtually impossible to accurately gauge the relative and absolute sums devoted to this effort. Whatever the precise figure, the perception in Britain persists that "health promotion activities receive a low priority in the allocation of constrained resources."[56]

Another problem in calculating and evaluating the actual commitment to disease prevention and health promotion is that it is sometimes difficult to differentiate between prevention and treatment, or to identify health promotion services. Certainly the NHS physician who in the course of examining a patient with pneumonia urges that person to quit smoking is engaged in an effort at life-style modification. Yet he or she will be reimbursed for treating an illness, not for promoting health.

There is, however, another way to evaluate the role of disease prevention in British health policy, by determining the extent to which the recommendations contained in the 1977 British White Paper have been achieved. David Cohen and John Henderson have done this in an article whose title, "No Strategy for Prevention," reveals its overall conclusion. The authors found that although both Labor and Conservative governments have substantially increased spending on health education—from £1.5 million ($2.6 million) in 1977 to £6.5 million ($11.4 million) in 1982—there has been little significant achievement in health promotion areas such as smoking, alcohol consumption, fluoridation, and diet. Henderson and Cohen attribute much of the failure in implementing more aggressive health promotion policies to the influence of bureaucratic and economic interests. They found, for example, that despite recommendations to discourage alcohol consumption by keeping the real price of alcohol in line with other products that "once again, as with smoking, it looks as though the Treasury's self-interest, in alliance with the relevant producers and manufacturers, has combined to thwart the letter and the spirit of the recommendations."[57] The overall conclusion concerning implementation of the health promotion recommendations of the White Pa-

[56] Madeline Rendall and Bobbie Jacobson, "Health Promotion in England, Wales and Northern Ireland," in Christopher Robbins, *Health Promotion in North America: Implications for the UK* (London: Health Education Council and King Edward's Hospital Fund, 1987), 17.

[57] John Henderson and David Cohen, "No Strategy for Prevention," in *Health Care UK 1984*, ed. Anthony Harrison and John Gretton (London: Chartered Institute of Finance and Accountancy, et al., 1985), 66.

per was that "even those recommendations which the Government accepted are making, at best, but slow progress."[58]

In addition to bureaucratic and economic self-interests, Rudolf Klein offers another explanation for the slow progress of disease prevention in British health policy. According to Klein, prevention has no constituency and its benefits are too abstract: "Those who will benefit cannot be identified; moreover, the benefit itself is uncertain. For prevention is about the reduction of statistical risk, not about the delivery of certain benefits to specific individuals."[59] These reasons, along with the specific issues mentioned earlier, such as concerns for individual freedom and scientific uncertainty, have resulted in high rhetorical, but relatively low legislative, priority for health promotion in Britain.

No discussion of health policy in Britain would be complete without recognition of the important role played by the professional health community. In subsequent chapters I will identify specific instances of active support for, or indeed initiation of, health promotion policies. Particularly noteworthy has been the almost militant position of the BMA on various life-style issues.

THE UNITED STATES: HEALTHY PEOPLE

In the mid-1970s, American political leaders shared the concern of their British counterparts with exploding health care costs and the potential benefits of a strategy of health promotion. In 1976 Congress enacted the National Consumer Health Information and Health Promotion Act (PL 94–317).[60] The law had two parts. The first dealt with traditional public health areas such as immunization of children and control of venereal and other infectious or communicable diseases. The second, and for my purpose the most important and innovative part of the law, was the allocation of funds for the creation of an office of Health Information and Health Promotion in the Department of Health, Education, and Welfare (now Health and Human Services).

The law recognized the importance of more health-enhancing life-styles among the American people, and the willingness of the government to encourage healthier habits. The initial allocation for health promotion was quite modest: $7 million in fiscal 1977, $10 million in 1978, and $14 million in 1979. The total of $31 million represented a substantial decrease from the original House version of $70 million and the Senate version of $49 million. The lowered figure was needed to overcome objec-

[58] Anthony Harrison and John Gretton, "Audit," in *Health Care UK 1984*, 48.
[59] Rudolf Klein, *The Politics*, 173.
[60] U.S. Public Law 94–317.

tions by the Ford administration to the health information and promotion section of the bill. The administration argued that this effort represented a duplication of services already performed by existing federal agencies and programs. The final version also differed from the original congressional bill in that it dropped a proposed new private health education center that was to be partially funded by the federal government.

The key provisions of the act required the secretary of health, education and welfare to

1. Develop a strategy of health education and promotion that might include research, information, and training programs, and granting funds to private, nonprofit organizations working in the area of health promotion.
2. Conduct periodic surveys of the health care needs and knowledge of Americans.
3. Report to Congress within two years on health promotion activities.
4. Determine the extent to which health insurance programs cover health promotion and education services.
5. Establish an Office of Health Information and Health Promotion to help coordinate these and related activities. (This was later renamed the Office of Health Information, Health Promotion, and Physical Fitness and Sports Medicine.)[61]

With the passage of PL 94–317, the federal government took cognizance of the relationship between life-style and ill health and committed itself, largely through health education, to improve individual health consciousness and habits.

A direct outgrowth of PL 94–317 was the 1979 surgeon general's report on health promotion and disease prevention, entitled *Healthy People*. The purpose of the report was "to encourage a second public health revolution." The document, according to Secretary of Health, Education, and Welfare Joseph Califano, represented "an emerging consensus among scientists and the health community that the Nation's health strategy must be dramatically recast to emphasize the prevention of disease."[62] *Healthy People* identified major health risks and specified health goals to prevent disease at each of the principal stages of life: infancy, childhood, adolescence, adulthood, and older adulthood. It suggested general "actions for health" to achieve these goals, including preventive health services, and health protection and health promotion activities. In the main, the document was a health promotion agenda, an exhortation to individuals and the business community for greater awareness and responsibility in promoting healthier behavior.

[61] See 1976 *Congressional Quarterly Almanac*, 538–40.

[62] United States Department of Health, Education, and Welfare, *Healthy People*, vii.

In 1980 the U.S. Public Health Service followed up on *Healthy People* with *Promoting Health/Preventing Disease: Objectives for the Nation.* The document suggested steps that could be taken by both the private and public sectors to reduce morbidity, disability, and premature mortality. In addition to suggesting traditional health promotion programs, such as immunization, occupational safety, and pre- and postnatal care, it covered the new concern with environmental health hazards and various life-style-related areas. There were sections on sexually transmitted diseases, fluoridation, smoking, alcohol and drug misuse, nutrition, physical fitness, and exercise. In each section of the report various prevention or promotion measures were offered to national, state, and local governments, and the private sector. In all 226 specific health objectives were established for 1990. The motivation behind the document, aside from the legislative mandate, was made clear in its introduction: "At a time in the Nation's history when budgets become tighter, legislators, public officials and governing boards of industry, foundations, universities and voluntary agencies are beginning to re-examine their traditional bases for allocating their limited health-related resources."[63]

In 1986 the Public Health Service issued a "midcourse review" on the nation's progress toward achieving the goals set by the original plan. The report predicted that 110 of the 226 objectives would be met by 1990 or had already been achieved. Included in these were the goals set for reductions in alcohol and cigarette consumption and motor-vehicle deaths. In the latter case progress was already substantial. Between 1978 and 1984 the automobile-accident death rate declined from 23.6 per 100,000 people to 19.6 per 100,000 (19.2 in 1985); the 1990 goal was 18 per 100,000. In addition, deaths from cirrhosis were down from 13.5 per 100,000 in 1978 to 11.6 in 1984 (11.2 in 1985); the 1990 goal had been 12 per 100,000. The review attributed this reduction to changes in life-style, including greater use of seat belts and more prudent use of alcohol before driving. The review was less optimistic with regard to other life-style patterns; there was not much change in the number of adults overweight or those who exercised regularly.[64]

The Health Promotion and Health Information Act, along with the surgeon general's report and *Promoting Health/Preventing Disease*, provides a clear statement of national recognition and purpose on life-style and disease. The emphasis has been on education and increasing national awareness and sensitivity to the causes and consequences of imprudent

[63] United States Department of Health and Human Services, *Promoting Health/Preventing Disease: Objectives for the Nation* (Washington, D.C.: Government Printing Office, 1980), vii.

[64] See *The 1990 Health Objectives for the Nation: A Midcourse Review* (Washington, D.C.: Public Health Service, 1986).

daily life-styles. In this regard the government periodically publishes studies aimed at alerting the public to the causes of volitional illnesses. The best known of these are the surgeon general's reports on smoking and health, but there have also been publications on AIDS and, in July of 1988, the first national Report on Nutrition and Health. It is clear that under both Democratic and Republican administrations, although a federal role is acknowledged, emphasis is placed first on the individual, and second on collective action by the private sector and state and local governments.

One final point about the federal role in this area. In its broadest definition health promotion and disease prevention include occupational and environmental safety, infectious disease control, maternal and child health and nutrition, and other programs. The federal government supports these efforts through Health Block Grants to the states, as well as the activities of the Centers for Disease Control, the U.S. Department of Agriculture, the Occupational Safety and Health Administration, and the Environmental Protection Agency programs. Thus, the federal government has undertaken, in the last decade or so, a wide-ranging involvement in the general area of health promotion and disease prevention.

In both the United States and Britain the decades of the 1970s and 1980s have witnessed increased acceptance of the need for greater individual responsibility in health promotion and disease prevention. In both countries the emphasis has been on educating people about those facets of their lives that increase the risk of illness, disability, and premature death. This second public health revolution (or "new perspective") has been in response to seemingly uncontrollable increases in health care costs, the changing nature of disease patterns, the limits of modern medicine, and the changes in society since World War II. In this latter category the changes in sexual norms and gender roles have been especially important. Exactly how people interpret these social changes and relate them to health issues depends on cultural and ideological preferences. Many conservatives, for example, view the AIDS epidemic and the dramatic increase in drug abuse as a confirmation of what happens when societies abandon traditional moral values and practices. What tends to happen in these and other areas of public health concern, then, is that health policy becomes a soldier in the war not only to reduce human misery, but also to save money, souls, and social values.

Despite the philosophical preference in both countries to promote prudent behavior through education, rather than regulation or coercion, in a number of instances it appeared that many in America and Britain were unwilling or unable to summon up the discipline to change injurious life-styles. In addition, many remained unconvinced that such sacrifices were either necessary or the business of government at all. Yet the costs of personal imprudence, its negative externalities, continued to burden the collective resources and stability of society. As a result, on several occasions public policymakers turned to regulation of reputedly risky personal behavior in the name of the collective good. The following chapters examine some of these efforts.

Four

Smoking and Health Policy: A New Prohibition?

IN FEBRUARY 1987 U.S. physicians in an organization called "Doctors Ought to Care" followed the lead of some of their British colleagues and began to send black-bordered death notices to members of Congress. The notices informed the legislators that one of their constituents had died of a smoking-related disease. The form included the name of the constituent, the specific disease, and an appeal to the legislator to oppose any bills aiding the tobacco industry. The death notice, which was characterized by a representative of the tobacco industry as "grotesque," reflects the emotional intensity generated by the issue of smoking and health. One American anti-smoking advocate has said, "Only the unquantifiable threat of nuclear annihilation poses a greater threat to health and life [than tobacco]."[1] A British counterpart referred to cigarettes as "the most lethal instrument devised by man for use in peacetime."[2] Terms such as "holocaust," "cancer club," "murderers," "retailers of disease," and "merchants of death" are commonplace in descriptions of cigarettes and the industry that produces and sells them.

Cigarette smoking occupies a unique place in the debate over life-style and health for three reasons. First, it is generally accepted by experts on both sides of the Atlantic that cigarette smoking poses the most serious public health problem of our time, comparable in consequence to the great epidemics of the nineteenth century. Smoking accounts for between 15 and 20 percent of all deaths in both countries annually, approximately 100,000 deaths in Great Britain (177 per 100,000) and 390,000 (163 per 100,000) in the United States. In the case of the United States, more people die prematurely each year from smoking-related diseases than from automobile and fire accidents, drug abuse, homicide, suicide, and AIDS *combined.* In terms of the alleged scope and breadth of its devastation, no other health-related personal choice comes close to cigarette smoking.

[1] R. T. Ravenholt, "Tobacco's Impact on Twentieth-Century Mortality Patterns," *American Journal of Preventive Medicine* 1, no. 4 (1985): 14–15.

[2] Sir George Godber quoted in Paul Castle, "Tobacco's Million Victims," *Health and Social Service Journal* 93 (18 August 1983), 993.

Second, cigarette smoking differs from the consumption of other allegedly injurious products, such as alcohol and foods high in sugar, salt, or cholesterol, in that health authorities agree that there is no such thing as a safe level of use: the product itself, not simply its abuse or misuse, constitutes the health problem. Finally, smoking differs from other health problems in the magnitude of behavioral change associated with it. Thirty years ago roughly one-half of the adult populations of the United States and Britain smoked cigarettes; today 29 percent of Americans and 33 percent of the British smoke. Thomas Schelling, of Harvard's Institute on Smoking Policy and Behavior, calls this change "one of the most spectacular social phenomena in the postwar period."[3]

Since the first authoritative reports linking smoking and various diseases appeared twenty-five years ago, the question of the appropriate role of government in controlling cigarette production and sales, and influencing the individual's decision to smoke, has been the substance of the most important public health debate of the period. In the following pages I will first describe the dimensions of the public health problem associated with smoking. Then I will discuss the similarities and differences in British and American policy approaches to the issue of smoking and health. What will emerge is the image of a generally more aggressive and coercive approach to discourage cigarette use in the United States than in Britain. Both countries began with health education to alert people to the dangers posed by smoking. As the evidence linking smoking and ill health became more voluminous and compelling, U.S. anti-smoking forces have been able to bypass a strong Congressional tobacco lobby and convince state and local lawmakers to move beyond education to more intrusive and freedom-limiting ways of modifying behavior.

In Britain, where there are no alternative policy-making arenas, and where Thatcherism prevails, education rather than regulation continues to dominate. The one weapon in the anti-smoking arsenal that the British have used with greater alacrity than American lawmakers has been taxation, although this may be changing. Taxes comprise a much higher percentage of the per-package cost of cigarettes in Britain than in the United States, and they have been rising more rapidly and consistently. Most importantly, in recent years these increases have been specifically targeted at decreasing consumption. Because taxation is a more oblique approach to this particular life-style change than, say, limiting smoking in public places, British policy has been perceived as less of a threat to individual freedom than the U.S. policy approach.

[3] Quoted in *Congressional Record*, (17 July 1986), 9315.

Dimensions of the Problem

Trends in Consumption

It is easy to forget how relatively recent the widespread use of cigarettes is. The modern history of the tobacco industry begins in 1884, when W. Duke and Sons of Durham, North Carolina—founders of the American Tobacco Company—introduced the first commercial cigarette-rolling machine. Even then, despite this technological breakthrough, cigarette consumption remained relatively modest through the first decade of the twentieth century. World War I, and the development of better-burning tobacco, stimulated a dramatic increase in tobacco consumption. The U.S. Expeditionary Force received free cigarettes, and General John J. Pershing told leaders in Washington, "Tobacco is as indispensable as the daily ration; we must have thousands of tons of it without delay."[4] Exactly how many men were first introduced to smoking because of the war is not clear. What is clear is that annual consumption increased from less than fifty cigarettes per adult in 1900 to nearly seven hundred and fifty in 1918. The situation in Britain was much the same, with per-adult consumption increasing from less than one cigarette per day in 1900 to over six in 1918 (see Figures 4-1 and 4-2).

The next few decades witnessed a steady growth in the consumption of cigarettes, although the proportion of the smoking population remained relatively small. Between 1920 and 1940, annual consumption in the United States increased from approximately six hundred to eighteen hundred cigarettes per adult, although the proportion of the population that smoked grew more moderately, from 15 percent to 20 percent. Part of the explanation for the small proportional growth in the adult smoking population is that prior to World War II, few women in either country smoked. This pattern changed following, and partly as a result of, the war.

The Second World War had an important impact on the image and consumption of cigarettes. A positive, indeed heroic and patriotic, picture of smoking was portrayed in the movies and the media. Soldiers and civilians, including an increasing number of women, found in smoking a way of alleviating their anxieties, and both the British and American governments supplied cigarettes to the military as part of the war effort. President Roosevelt, who was himself often photographed smoking, declared tobacco an essential commodity and granted draft deferments to tobacco

[4] Quoted in Susan Wagner, *Cigarette Country: Tobacco In American History and Politics* (New York: Praeger Publishers, 1971), 44.

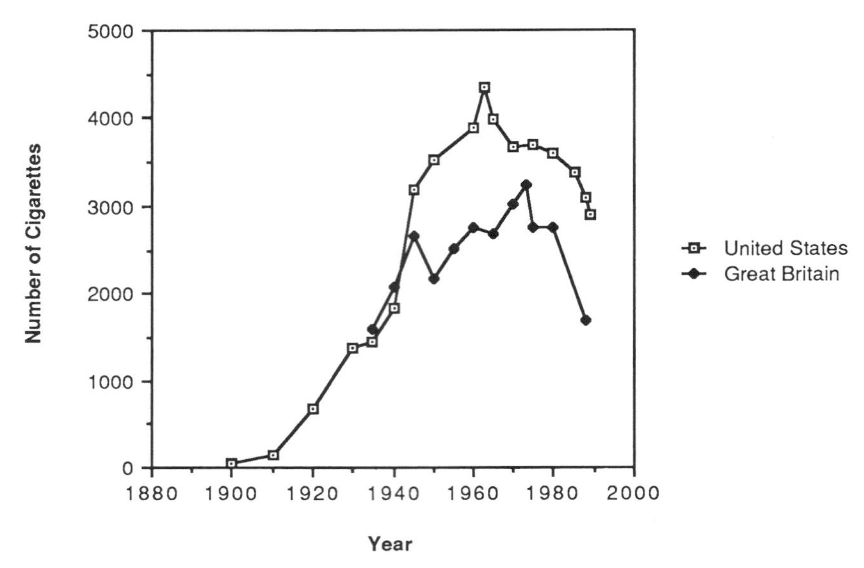

FIGURE 4-1

Per Capita Cigarette Consumption, United States and Great Britain, 1900–1988

Sources: United States Department of Health and Human Services, *Reducing the Health Consequences of Smoking: Twenty-Five Years of Progress*, report of the Surgeon General (Washington, D.C.: Government Printing Office, 1989), 268; *Oregonian*, 29 April 1990. Royal College of Physicians, *Health or Smoking?* (London: Pitman Publishing, 1983), 1; *Economist* 26 August 1989, 54.

farmers. At home and overseas, cigarette smoking became the habit of a significant proportion of the adult population. In 1940, 20 percent of the adult population in the United States were smokers; by 1949 the proportion had more than doubled to 44 percent. Annual cigarette consumption rose from 1,828 cigarettes per adult in 1940 to 3,200 by 1945. Although prewar data on the percentage of adults smoking is not available for Britain, consumption patterns paralleled the trends in the United States: annual per adult consumption increased from 1,590 cigarettes in 1935 to 2,665 in 1945.

The proportion of smokers stabilized by the early 1950s at around 42 percent in the United States and 55 percent in Britain, but consumption continued to rise. In the United States increased consumption did not peak until 1967, and in Britain not until 1973. The upward trend in consumption was finally halted by the accumulation of evidence linking smoking and ill health.[5]

[5] In 1987 U.S. Surgeon General C. Everett Koop noted that there were 50,000 studies linking smoking to various diseases. See the *New York Times*, 15 March 1987.

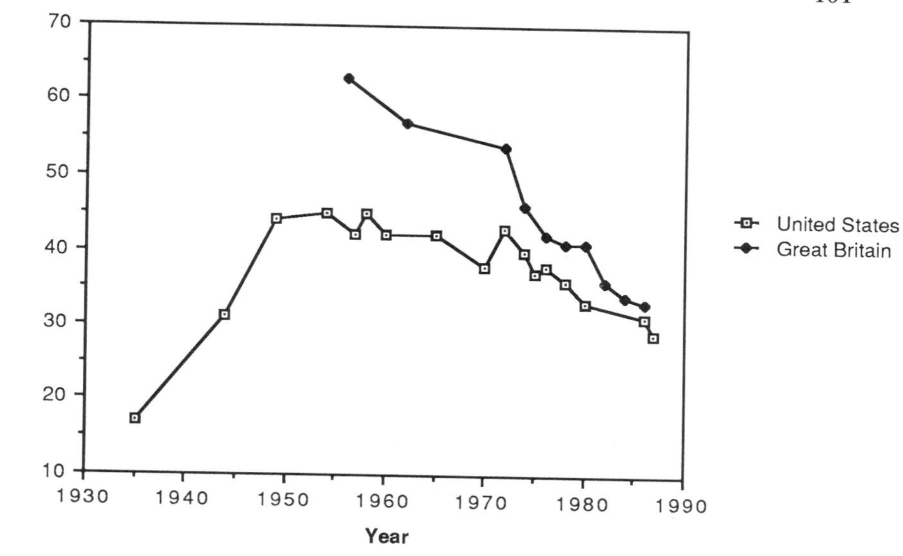

FIGURE 4-2
Cigarette Smoking, Percentage of Adult Population, United States and Great Britain, 1935–1987

Sources: United States Department of Health and Human Services, *Reducing the Health Consequences of Smoking: Twenty-Five Years of Progress*, report of the Surgeon General (Washington, D.C.: Government Printing Office, 1989), 269. Royal College of Physicians, *Health or Smoking?* (London: Pitman Publishing, 1983), 107; Great Britain, Central Statistical Office, *Social Trends 19* (London: Her Majesty's Stationery Office, 1989), 123.

Smoking and Illness

Concerns about potential adverse health effects of tobacco have been raised for centuries, but prevailing lay and scientific wisdom was that smoking was a pleasurable, indeed beneficial, practice. This position was endorsed by no less an authority than the U.S. Supreme Court in a case in 1900 (*Austin v. Tennessee*) dealing with a Tennessee state law prohibiting the sale of cigarettes. The Court noted that tobacco's "extensive use over practically the entire globe is a remarkable tribute to its popularity and value." Failure to recognize the dangers of smoking was not limited, however, to jurists. A 1948 editorial in *JAMA* proclaimed that, "there does not seem to be any preponderance of evidence that would indicate [that tobacco is] a substance contrary to the public health."[6]

Despite assertions of confidence in the benefits derived from tobacco

[6] Quoted in *Congressional Record* (17 July 1986), 9316.

use, scientific evidence began to point in quite the opposite direction. In the 1920s and 1930s a number of empirical studies in the United States, Britain, and Germany suggested links among cigarette smoking, lung cancer, and decreased longevity.[7] But it was not until the late 1940s and early 1950s that some of this work began to attract attention beyond esoteric medical and scientific journals. These studies were prompted in part by epidemiological and impressionistic evidence in the medical community of a dramatic increase in the mortality rate from lung cancer.[8]

In the late 1940s researchers in the United States and Britain began retrospective studies involving lung cancer victims. The studies, and their astonishing results, were first published in the early 1950s. In a May 1950 article in *JAMA*, Doctors Ernest Wynder and Evarts Graham reported that 95 percent of the 650 men with lung cancer they studied had smoked for at least twenty years. Four years later Doctors E. Cuyler Hammond and Daniel Horn of the American Cancer Society reported "that men smoking a pack or more a day were five times as likely to die of lung cancer as nonsmokers and twice as likely to die of heart disease."[9] These and other studies led the Tobacco and Cancer Committee of the American Cancer Society to adopt a resolution in 1954 recognizing a link between cigarette smoking and lung cancer.

The mounting scientific evidence was too strong for policymakers to ignore. In February 1954 the British cabinet was informed of the evidence linking smoking and lung cancer in a report from an official committee chaired by the Government Actuary. The report concluded that "a real association between smoking and cancer of the lung was firmly established; and the connection was causal."[10] (The report was not made public until 1985 under the thirty-year rule for official documents.) Despite this conclusion, when the minister of health, Mr. Ian McLeod, informed the House of Commons of the findings, he presented a less ominous characterization of the danger involved. In his report to the Commons on February 12, 1954, Mr. McLeod noted, "I would draw attention to the fact that there is so far no firm evidence of the way in which smoking may cause lung cancer or of the extent to which it does so. We must look to the results of research and its vigorous pursuit to determine further action."[11] (Some years later Mr. McLeod provided some insight into gov-

[7] For citation of these studies see United States Surgeon General, *The Health Consequences of Smoking: Cancer* (Washington, D.C.: Department of Health and Human Services, 1982), 3; Wagner, *Cigarette Country*, 68–69.

[8] See Ravenholt, "Tobacco's Impact," 6; and Royal College of Physicians, *Health or Smoking?* (London: Pitman Publishing, 1983).

[9] Quoted in *Congressional Record* (17 July 1986), 9316.

[10] See *Times* (London), 7 January 1985.

[11] Ibid.

ernment attitudes toward cigarettes when he said, "Smokers, mainly cigarette smokers, contribute some £1,000 million yearly to the Exchequer and no one knows better than the government that they simply cannot afford to lose that much."[12])

The first official acknowledgment of the health danger in the United States came from the surgeon general, Leroy E. Burney, in July 1957. It was not, however, until the early 1960s that the issue of smoking and health became firmly established as a national political issue. In both countries the catalyst in the transformation of this from a peripheral to a central political issue was the publication of a report on smoking and health by two highly respected health authorities. The first of these was the 1962 publication of the British Royal College of Physicians' report, *Smoking and Health*; the second, the 1964 U.S. surgeon general's report of the same title.

Since 1725, when it alerted the House of Commons of the dangers of cheap gin to the public health, the Royal College has taken a leading role in a number of controversial public health issues. In 1962 the College addressed the question of whether cigarette smoking posed a threat to the public's health. In its report *Smoking and Health* it concluded that "cigarette smoking is the most likely cause of the recent world-wide increase in deaths from lung cancer . . . ; that it is an important predisposing cause of the development of chronic bronchitis [and it] increases the risk of dying from coronary heart disease."[13] The College urged that "decisive steps" be taken to curb smoking, including limiting smoking in public places, curtailing advertising, increasing cigarette taxes, and mounting a vigorous health education campaign. These recommendations provided the British smoking and health policy agenda for the next twenty-five years.

Encouraged by the efforts of their British colleagues, the U.S. health community prodded the Kennedy administration to undertake its own investigation into the relationship between smoking and ill health. Although reluctant to antagonize state tobacco interests, and the enormously influential congressional tobacco lobby, President Kennedy nevertheless approved the appointment of an Advisory Committee on Smoking and Health. One measure of the tobacco lobby's influence was that potential committee members were scrutinized by, and subject to the veto of, the Tobacco Institute, an industry public relations and lobbying organization. In January 1964 the committee reported on its findings in *Smoking and Health*; the news was not good for the tobacco industry.

[12] Peter Taylor, *Smoke Ring: The Politics of Tobacco* (London: The Bodley Head, 1984), 5.

[13] Royal College of Physicians, *Smoking and Health* (London: Pitman Medical and Scientific Publishing Company, 1962), 43.

The report concluded that cigarette smoking was causally related to lung cancer, strongly associated with other cancers such as cancer of the mouth and larynx, and the most important cause of chronic bronchitis.

The two reports were benchmarks in the debate on smoking and health. Over the next twenty-five years literally thousands of scientific papers and reports on the health consequences of smoking, including three more Royal College reports and seventeen additional studies by the U.S. Public Health Service, would expand the indictment of cigarette smoking. In 1982 U.S. Surgeon General C. Everett Koop called cigarette smoking "the chief, single, avoidable cause of death in our society and the most important public health issue of our time."[14]

The public health picture that emerged from these reports was one of doom and gloom. By one estimate, more than 10 million people in the United States have died from tobacco-related illnesses in this century alone.[15] Lung cancer in particular has attracted attention and concern because of its virtually undisputed relationship to cigarette smoking. In the United States, lung cancer deaths, which were virtually unknown in 1900, have gradually increased over the century—3,000 (2 per 100,000 population) in 1930, 18,000 (12 per 100,000 population) in 1950, 40,000 (21 per 100,000 population) in 1964, and 111,000 (48 per 100,000 population) in 1982. Today lung cancer is the most common form of malignancy among both men and women. In Britain, which has among the worst lung cancer rates in the world, the pattern has been the same: the average annual death rate among men aged 45–64 from lung cancer between 1916 and 1920 was 146 per 100,000 population; by 1956–1959 it was 9,108 per 100,000 population.[16]

It is difficult to place a precise price tag on the costs of cigarette smoking, but it is, by most accounts, considerable. In a report to Congress in February 1990 the United States secretary of health and human services estimated that the annual cost, including health care costs linked to smoking, insurance, and lost economic productivity, was $52 billion or about $221 per person. The anti-smoking group Action on Smoking and Health challenged this estimate, suggesting the cost was closer to $100 billion per year.[17] In Great Britain it is estimated that the annual cost to the National Health Service alone is £500 million ($320 million).

[14] United States Surgeon General, *The Health Consequences of Smoking: Cancer*, xi.

[15] Ravenholt, "Tobacco's Impact," 14.

[16] See Pascal James Imperato and Greg Mitchell, *Acceptable Risks* (New York: Viking, 1985), 21; Ravenholt, "Tobacco's Impact," 12; *Times* (London), 5 December 1982; *Congressional Record*, (21 July 1983), 3650.

[17] *New York Times*, 21 February 1990.

Smoking, Health, and Other Social Agenda Concerns

Over the years the smoking and health debate has become more particularized as the impact of smoking on certain segments of British and American society is scrutinized more closely. Recently the issue has provided an arena for debates over other social agenda issues, including the problems facing women, minorities, and the poor in each of these countries.

The relationship between race and smoking has only just emerged in the overall smoking and health debate in the United States. Much of the initial public health concern stems from the fact that blacks have the highest mortality and morbidity rates from the major smoking-related diseases—lung cancer and coronary heart disease—of any group in society. In 1983 the age-adjusted lung cancer death rate for black males was 44 percent higher than for white males, while the heart disease rate was 20 percent higher (308.2 versus 257.8 per 100,000 population).[18] There may be a variety of factors that explain these higher rates, such as higher serum cholesterol levels and greater prevalence of overweight among blacks, but at least part of the problem stems from the higher incidence of smoking among blacks: a 1987 Gallup survey found that 39 percent of blacks polled compared to 30 percent of whites smoked. Another study revealed that the quitting rate was lower among blacks (24.9 percent) than whites (39 percent).[19] It should be noted that much of this difference is a function of lower socioeconomic status and education level.

Antismoking groups, including Doctors Ought to Care, have charged that tobacco companies are particularly targeting blacks in their advertising and promotional campaigns. One survey found that there were three times as many billboards in St. Louis's black neighborhoods as in white ones, and that 62 percent of the billboards in the black neighborhoods advertised cigarettes and alcohol compared to 36 percent in the white neighborhoods.[20] In early 1990 R. J. Reynolds Tobacco Company sparked a storm of protest by announcing its intention to test-market a new menthol cigarette, "Uptown," specifically aimed at the black population. Although Reynolds denied any impropriety in group targeting, it was forced to cancel the campaign after being publicly rebuked by the secretary of health and human services, the American Cancer Society, and the National Association for the Advancement of Colored People

[18] National Center for Health Statistics, *Health, United States, 1985* (Washington, D.C.: Government Printing Office, 1985).

[19] *New York Times*, 17 January 1987; *Gallup Report* No. 258; *New York Times*, 17 January 1987.

[20] *New York Times*, 1 May 1989.

(NAACP).[21] As more and more white middle-class males abandon smoking, tobacco companies have sought to extend their market to proven reliable consumers like blacks and the working class, or to tap not fully exploited markets like young people and women. In addition, Dr. Alan Blum, founder of Doctors Ought to Care, has accused the tobacco industry of buying off black leaders and black publishers through gifts to black organizations and causes and heavy advertising in black publications.

Black leaders have acknowledged that anti-smoking campaigns have not played a prominent role in their effort but have defended the support given to black causes and black publications by tobacco companies. James Williams of the NAACP has asked, "What's more important: that the United Negro College Fund receives hundreds of thousands of dollars in contributions from R. J. Reynolds for scholarships or that it advertises in one of the fund's publications?" And Benjamin Hooks, Jr., executive director of the NAACP, sees the attack on the relationship between blacks and the tobacco industry as another form of racism. "If black leaders ask for funding from the tobacco industry we're accused of 'selling out.' Whites get billions of dollars from these same companies and for some unknown reason that's not viewed as a sellout."[22] For its part, the tobacco industry rejects the notion that targeting special groups is malevolent; it is a standard advertising practice. Furthermore, the Tobacco Institute has charged that those who attack such advertising are "at the very least paternalistic, or even offensive, in suggesting that blacks are too stupid or obtuse to see a product that is legally advertised and make their own decisions."[23]

The position of black leaders on the issue of smoking and health in the United States is somewhat analogous to that of labor leaders in Britain. Working-class people, especially those in the lowest socioeconomic classes, are more likely to smoke, less likely to quit, and more likely to die prematurely from smoking-related diseases than those in the higher social classes. In 1984, 49 percent of male "unskilled manual laborers" smoked cigarettes compared to just 17 percent of male "professionals." Despite the disproportionate health burden on the poor, trade unions and Labor party leaders traditionally have been reluctant to impose anti-smoking policies for fear of the adverse impact on one of the "workingman's few pleasures." Although he thinks this attitude is changing, David Simpson of Action on Smoking and Health charges that "the Labour Party in this country has been very wary of touching smoking too hard because they see it as affecting more of their voters."[24] In addition, trade union leaders

[21] *New York Times*, 12 January 1990.
[22] *New York Times*, 17 February 1987.
[23] Portland *Oregonian*, 7 February 1988.
[24] David Simpson, interview by author. Tape recording, London, England, 30 May 1984.

have not been unmindful of the employment implications of reduced smoking for their membership. For a number of years the annual meeting of the Trades Union Congress, the umbrella organization for many of Britain's unions, grappled with establishing a position on various smoking issues, including opposition to cigarette advertising and sports sponsorship, and a smoke-free environment in the workplace. In these efforts, however, the Trades Union Congress met with opposition from members of the Tobacco Workers Union. (Much the same can be said about American labor unions.)

Recently attention has also been focused on the issue of women and smoking. In 1980 the surgeon general devoted his annual report on smoking and health to the health consequences of smoking for women, and the most recent Royal College of Physicians report, "Health or Smoking?" includes a chapter on women and smoking. The Royal College report noted that "the recognition that smoking in women is an important issue is perhaps one of the major developments in smoking research over the last decade."[25] The recent concern over the impact of smoking on women's health is all the more interesting because it contrasts with the historical inattention accorded to the problem. This neglect, on the part of both policymakers and the medical community, was the result of the significantly lower rate of smoking-related illnesses among women. For example, in Britain in 1969 there was only one female death due to lung cancer for every five male deaths. Similarly, in the United States in 1950 women accounted for less than one out of every twelve lung cancer deaths. This epidemiological pattern led many to "the fallacy of women's immunity" from smoking-related illnesses.[26]

By the beginning of the 1980s, however, the notion of female immunity was discredited. By 1980 three women in Britain died from lung cancer for every five males; deaths from lung cancer rose from 5,000 in 1969 to 8,400 in 1980. In the United States by 1980, over one-quarter of all lung cancer deaths occurred in women compared to less than ten percent in 1950. Furthermore, in both countries the rate of increase in lung cancer is higher in women than men, and lung cancer has replaced breast cancer as the number-one cause of death among women. The recent "epidemic of lung cancer" and other smoking-related illnesses among women is explained by the fact that women trailed men by more than twenty-five years in adopting the smoking habit. As the surgeon general's report noted, "By the time of the first retrospective studies of smoking and lung

[25] Royal College of Physicians, *Smoking or Health?* (London: Pitman Medical Publishing Company, 1977), 61.

[26] See both Royal College of Physicians, *Health or Smoking*, and United States Surgeon General, *The Health Consequences of Smoking for Women* (Washington, D.C.: Department of Health and Human Services, 1980).

cancer in 1950, two entire generations of men had already become lifelong cigarette smokers. Relatively few women from these generations smoked cigarettes, and even fewer had smoked cigarettes since their adolescence."[27] With the post-World War II increase in cigarette use among women—by 1960 34 percent of U.S. women and 43 percent of British women were smokers—the lung cancer rate began to increase.

The increased proportion of women who smoke, and the increased amount they smoke, was probably the result of a variety of psychological, social, and behavioral factors associated with the changing role of women in society. Research on the reasons for increased smoking among women has produced no definitive answers, but it suggests that smoking is used by women more than men during periods of "high arousal" (emotional strain, anxiety, and demanding mental activity) and "negative affect." Recent research indicates that the conditions producing these situations particularly afflict women in the workplace. Such conditions include "discrimination against women in employment, role conflict, authority problems, inequity in promotions, exclusion from decision-making processes and the 'old boys' network." In addition, it has been suggested that the tensions produced by childrearing and working have also contributed to increased cigarette consumption among women.[28] The tentative conclusion one can reach from this research is that as women entered careers of high stress, previously closed to all but a few of them, and increasingly managed both careers and families, cigarettes helped ease the unprecedented psychological and social burdens attending these role changes.

With increased cigarette use among women came increased scientific attention to the unique health consequences of smoking for women. Epidemiologists paid particular attention to the relationship between smoking and reproduction. Both the recent Royal College and surgeon general reports endorse the research conclusions that smoking adversely affects the health of women who use oral contraceptives, female fertility, the development of the fetus, and normal pregnancies.[29]

The policy implications of these gender-related health problems have touched upon questions involving advertising and health education. As women became a larger part of the cigarette-consuming public, tobacco companies began targeting them as a group. Advertising campaigns ("You've come a long way baby"), cigarette brands (Eve, Virginia Slims, Satin), and promotional activities (the Virginia Slims tennis tournament)

[27] U.S. Surgeon General, *The Health Consequence of Smoking for Women*, vi.

[28] See Bobbie Jacobsen, *The Ladykillers: Why Smoking Is a Feminist Issue* (London: Pluto Press, 1981).

[29] See Royal College of Physicians, *Smoking or Health?*, chapter 7, and "Pregnancy and Infant Health," in United States Surgeon General, *The Health Consequences of Smoking for Women*.

have sought to portray female smokers as young, rich, attractive, physically active, and sophisticated.

Policy responses to this particular facet of the smoking and health problem have been rather tentative, although more so in the United States than in Britain. The British voluntary agreement on cigarette advertising of March 1986 contained a provision prohibiting advertisements in magazines specifically aimed at women (that is, with a female readership of over 200,000). In addition, as in the United States, Britain adopted rotating health warning messages, one of which says, "Smoking When Pregnant Can Injure Your Baby and Cause Premature Birth." In the United States no such restrictions have been placed on advertising in women's magazines. It would appear that publishers of women's magazines face the same dilemma as leaders of black organizations: cigarette advertising is an important source of revenue. Kenneth Warner recently summarized research on the influence of tobacco advertising on coverage of smoking and health in women's magazines and "found a total of eight feature articles from 1967 to 1979 that seriously discussed quitting or the dangers of smoking—less than one article per magazine for more than a decade." Warner reported specific cases in which *Harper's Bazaar*, *Family Circle*, and *Cosmopolitan* made decisions not to publish anti-smoking material for fear of the impact on cigarette advertising in their magazines.[30]

With these general political and social trends in mind I can now turn to the specifics of the debate on smoking and health, comparing the evolution of the policy response in each country.

Politics and the Debate on Smoking and Health

Wealth or Health?

The debate on smoking and health has been portrayed largely as a confrontation between the economic interests of those who earn some or all of their income from tobacco, and the health of the people. In its most succinct formulation the conflict is seen as one between "long-term health and the short-term economic interests involved in smoking."[31] Recently Peter Taylor, in a well-received comparative study of the politics of tobacco in Britain and the United States, paid tribute to these powerful economic interests: "The battle to break the Smoke Ring [i.e., all those who benefit financially from cigarettes] is a battle between wealth and

[30] Kenneth E. Warner, Cigarette Advertising and Media Coverage of Smoking and Health," *New England Journal of Medicine* 312 (February 7, 1985): 386. See also Davis, "Current Trends In Cigarette Advertising and Marketing," *New England Journal of Medicine* 316 (March 1987): 729.

[31] Lester Breslow, "Control of Cigarette Smoking from a Public Policy Perspective," *Annual Review of Public Health* 3 (1982): 130.

health."[32] Taylor, like scores of others, leaves no doubt as to who has been winning the battle thus far.

It would seem perverse—and be wrong—to reach any conclusion other than that the "Smoke Ring" has exerted extraordinary influence over the content of public health politics and policies in both countries. And, it is not at all surprising that it does so. Tobacco, especially in the form of cigarettes, is an important economic product, generating considerable wealth for members of the "Smoke Ring." Governments in both countries derive considerable sums from cigarette and other tobacco product taxes, although in the United States revenues from these taxes have been declining in relative terms. Nevertheless, they still generate billions of dollars for both the federal and state governments, and in Britain cigarette taxes are the third largest source of consumer revenue after the value-added and oil taxes. In addition, cigarettes, which are the most heavily advertised products in the United States, generate literally billions of dollars each year in revenues for the advertising, newspaper, and magazine industries. This, in turn, has had important implications for the relationship between the media and the tobacco industry. As Kenneth Warner has noted, "Studies dating back to the 1930s provide evidence that the media's dependence on revenue from cigarette advertising has repeatedly led to suppression of discussion of smoking and health matters."[33]

The influence of the tobacco industry—which generated $55 billion in retail sales in the United States in 1988—is exerted most directly by the major multinational tobacco companies, including RJR Nabisco (formerly R. J. Reynolds) and Philip Morris, which account for 70 percent of the U.S. market, and American Brands, British-American Tobacco, and the Imperial Group in Britain. In 1989 Philip Morris, with sales of over $39 billion, ranked seventh in the American Fortune 500, RJR Nabisco, with sales of over $15 billion, ranked twenty-fourth, and American Brands, with over $7 billion in sales, ranked sixty-fourth. British-American Tobacco ($14 billion in sales) ranked sixty-fifth in Fortune's International 500 in 1988.[34] Although all of these companies have diversified in the last decade, their commitment to and in tobacco remains considerable. Despite diversification, because of the extraordinarily high profits on cigarettes, income and earnings from tobacco continue to provide a majority of the profits earned by companies such as RJR Nabisco and Philip Morris.

The political influence of the tobacco companies is often articulated, facilitated, or championed by public officials from tobacco-producing or

[32] Taylor, *Smoke Ring*, 274.

[33] Warner, "Cigarette Advertising," 385.

[34] See *Fortune* 121 (23 April 1990), and *Fortune* 120 (31 July 1989).

manufacturing areas. In the United States, this largely involves North Carolina and Kentucky, which combined produce about 63 percent of the nation's tobacco, as well as Tennessee, Virginia, South Carolina, and Georgia. Legislators from these states have been reminding their colleagues for decades of the vital role played by tobacco in the nation's economic and fiscal well-being. This role currently includes providing jobs and income for about 275,000 tobacco-growing farm families and 1.6 million others who earn a living in whole or part from the manufacturing, sales, or promotion of tobacco products; generating billions of dollars in public revenues; and providing about 2.5 billion dollars in export earnings in 1988.[35]

The influence of the congressional tobacco lobby would be limited if it involved only these few states: tobacco is grown in only fifty-one of the 435 Congressional districts, and in only twenty-seven of these is it the major economic enterprise.[36] The success of the tobacco-state congressional delegation in forestalling or delaying policies that would adversely affect the tobacco industry has been due in part to its ability to externalize the negative consequences of tampering with this industry. Many years ago former Senator Sam Ervin (D) of North Carolina outlined the purported ripple effect that anti-tobacco legislation would have. First, other agricultural-state senators were reminded of the impact that restrictive legislation might have on their own states' products: "When you destroy the growing of tobacco you are going to destroy every other farm program in the United States because these thousands of acres that are now devoted to the growing of tobacco will have to be devoted to the growing of sugarcane, sugar beets, cotton, and corn." Second, anti-smoking legislation would affect nonagricultural economic interests as well. "Not only would this bill have the effect of economic chaos on 625,000 farm families in America, but it would be well to note it would disable these farm families to buy automobiles from Michigan, and citrus products from California."[37] By building coalitions, especially with other agricultural interests, tobacco-state legislators have often delayed or modified anti-smoking legislation in Congress. In addition, tobacco-state legislators have been able to engage in traditional logrolling with nonagricultural-state lawmakers. Larry White reports that a proposed 1986 increase in the cigarette excise tax was defeated in the House Ways and Means Committee when New York City Congressman Charles B. Rangel (D) opposed it. According to White, "The ostensible reason was because excise taxes

[35] These points were detailed by Senator Sam Ervin (D-NC) in a speech defending tobacco subsidies. *Congressional Record* (15 July 1971), 25377.

[36] Larry C. White, *Merchants of Death: The American Tobacco Industry* (New York: Beech Tree Books, 1988), 48.

[37] *Congressional Record* (15 July 1971), 25377.

are regressive and fall most heavily on the poor. The real reason was that the North Carolina Democratic congressional delegation had cultivated Rangel and given him their votes on issues affecting his [primarily black] constituency in exchange for his support on tobacco."[38]

Finally, the tobacco lobby cultivates non-tobacco-state legislators in another way: in 1988 the Tobacco Institute paid more in speaking fees than any other interest group to members of Congress—$123,400 to eighty-five members of Congress. In Britain the influence of the tobacco industry is less visible and consequential at the legislative level than it is within the administration. It is true, to be sure, that about sixty members of Parliament represent constituencies dependent on tobacco manufacturing. Some of these M.P.s are "consultants" to tobacco companies and have been instrumental in preventing anti-smoking private members' bills. Nevertheless, the significant policy decisions are made by the prime minister and the cabinet, and it is at this level that the arguments about cigarette revenues, balance-of-payment deficits, and unemployment are most meaningful. Much of the political bargaining over policies affecting the tobacco industry occurs in an environment of noncoercive consultations between government ministries and industry representatives and results in voluntary agreements that effectively bypass the legislative process. This process has been alternately referred to as "tripartism," when it involves trade unions as well as industry and government, or "liberal corporatism" or "neocorporatism" to distinguish it from the more authoritarian, coercive corporatism of Latin America or Mussolini's Italy.[39] The reliance on cooperative consultations and agreements has long been characteristic of British industrial policy-making. David Vogel has contrasted the British inclination for more flexible, informal, and self-regulatory policy-making with the U.S. reliance on formal and detailed rules imposed by government. He credits the British approach, which he dates back to mid-Victorian times, to "a highly respected civil service, a business community that was prepared to defer to public authority, and a public that was not unduly suspicious of either the motives or power of industry."[40] Although Thatcher has largely abandoned neocorporatist administrative policy-making in most areas of economic regulation, it

[38] White, *Merchants of Death*, 231.

[39] See Richard Rose, *Politics in England* (Glenview, Illinois: Scott, Foresman and Company, 1989), 237–42; Dennis Kavanagh, *Thatcherism and British Politics: The End of Consensus?* (New York: Oxford University Press, 1987), 29; Raymond Plant, "The Resurgence of Ideology," in *Developments in British Politics*, ed. Henry Drucker, Patrick Dunleavy, Andrew Gamble, and Gillian Peele (New York: St. Martin's, 1983), 10–11; and Samuel H. Beer, *Britain against Itself: The Political Contradictions of Collectivism* (New York: Norton, 1982), 64–66.

[40] David Vogel, *National Styles of Regulation: Environmental Policy in Great Britain and the United States* (Ithaca: Cornell University Press, 1986), 26.

continues to prevail in issues affecting the tobacco industry, for example, advertising, sport sponsorship, and labeling.

In addition, some policy deliberations have traditionally involved primarily intragovernmental actors, and these in particular have demonstrated the political tension between advocates of health promotion on the one hand and the defenders of the public purse on the other. One particularly revealing illustration of this conflict involved a 1971 confidential government document on cigarette smoking and health. The report was prompted by the second Royal College of Physicians study, "Smoking and Health Now," and was a product of an interdepartmental group of officials comprised of representatives of the Treasury, Health and Social Security, Customs and Excise, and Employment Departments. The group considered the possible impact on cigarette consumption of new health warnings, a health education campaign, and increased cigarette taxes. It concluded that a decline in cigarette consumption, whatever its health implications, would have adverse consequences for the public treasury in terms of decreased revenues and increased social security benefits. The study group disbanded after failing to reach agreement on a government policy toward smoking. Furthermore, the *Guardian*, which first published the confidential report on May 6, 1980, claimed that "successive governments have conceded privately that although DHSS [Department of Health and Social Services] officials are keen on anti-smoking legislation, counter-arguments from the Treasury or the Department of Employment have always won the day." Although no direct evidence supports the contention that governments since 1971 have deliberately chosen to subordinate the health care needs of Britain for economic reasons, national economic considerations have undoubtedly had a profound impact on health policy-making. As the *Guardian* headline proclaimed when the report's contents were revealed, "Smoking not health won the day."[41]

As this example suggests, the influence of the tobacco industry need not be explicitly engaged. Chancellors of the Exchequer and ministers of trade need only consult revenue, export, and employment figures to be convinced of the importance of tobacco to the economic well-being of the nation: in 1986–1987 tobacco sales produced £5.6 billion ($8.2 billion) in taxes. Furthermore, approximately 21,500 persons are directly employed by the tobacco companies, and an additional 170,000 work in tobacco-dependent enterprises—tobacco shops, cigarette machinery manufacturing, filter-making plants, packaging companies, and so on. The sensitivity of the government to the economic implications of anti-smoking measures is heightened by the fact that much of the direct em-

[41] *Guardian*, 6 May 1980.

ployment in the tobacco industry is in already economically depressed areas of the country, including politically troubled Northern Ireland. Tobacco companies have periodically reminded the public, and their political leaders, of the relationship between policies that adversely affect the tobacco industry and unemployment.

It is then from a position of considerable economic strength that the British tobacco industry influences the course of public policy. This influence is most obvious during the periodic consultations between the government and the tobacco industry over the voluntary agreements relating to advertising, sales practices, and marketing. As one observer put it, "Voluntary agreements have given the tobacco companies the opportunity to shape the form of control through lengthy negotiations."[42] The tobacco industry's influence is enhanced by the decentralization of jurisdiction within the government over smoking and health policy. For instance, despite considerable opposition to sports sponsorship among civil servants in the DHSS, the tobacco industry has been able to avoid significant limitations in this area largely because it negotiates voluntary sponsorship agreements with the much more sympathetic members of the Ministry for Sport, who believe that the financial viability of certain minor sports such as badminton, golf, and snooker depends on such sponsorship.

The Evolution of Policy on Smoking and Health

Interestingly enough, some of the earliest efforts to address the problems associated with cigarette smoking were also the most restrictive. During the 1920s several U.S. states banned cigarette sales in the name of protecting public safety and morality. This "prohibitionist" movement was both temporally and philosophically coincidental with the other Prohibition effort, and its rise and demise paralleled that noble experiment; by 1927 all of the cigarette prohibition laws had been repealed.[43] It should be remembered that this was a time when the tobacco industry was only just emerging as a political and economic force.

It was not until the 1960s, with the publication of the authoritative reports of the British Royal College of Physicians and the U.S. surgeon general, that health considerations became a dominant issue in the public debate over the sale, promotion, and use of cigarettes. By this time, of course, cigarettes had an enormous and diverse constituency with about

[42] Christopher Hird, "Taking on the Tobacco Men," *New Statesmen* 27 (February 1981), 6.

[43] See United States Surgeon General, *The Health Consequences of Involuntary Smoking* (Washington, D.C.: Government Printing Office, 1986), 266–67.

one-half of the adult population smoking, and hundreds of thousands involved in growing, producing, selling, and advertising cigarettes. During the decades of the 1960s and 1970s, the effective beginning of the debate on smoking and health, the dominant issues in both countries involved how best to alert smokers to the putative dangers of cigarettes. More recently, beginning in the late 1970s, and accelerating in the 1980s, a sort of neoprohibitionist sentiment emerged in the United States, where health advocates have urged severe limitations on smoking in virtually all places but the smoker's own home. This second recent phase of the debate, which has had greater vitality in the United States than in Britain, has become a classic confrontation between public health and personal freedom.

Labeling and Advertising

In both Britain and the United States the first policy response to the published reports on smoking and disease was to alert smokers to the dangers of cigarettes. British and American governments required that warnings be placed on cigarette packages and imposed restrictions on some forms of cigarette advertising. Although both countries took basically the same approach, there were important differences between the two.

Health warnings have been required by law on cigarette packages in the United States since 1964. However, it was not until the 1970 Public Health Cigarette Smoking Act that Congress formally banned advertising of cigarettes on radio and television. The British, on the other hand, began in 1965 with a ban, through voluntary agreement, on advertisements of cigarettes on commercial television. This was followed in 1967 with a newspaper and magazine ban, and in 1968 with a prohibition of enticements, such as games, coupons, and other promotional efforts. It was not until 1971, nine years after the Royal College report, and six years after the advertising ban, that a health warning was placed on cigarette packages in Britain.

The difference in the policy sequence was partly fortuitous. The decision in Britain to begin a government anti-smoking campaign with a ban on television advertising was due in part to the fact that the government could act immediately without parliamentary approval or consultation with the industry. In 1964 Parliament had passed a Television Act renewing the licenses of the program contractors of the British commercial television system, the Independent Television Authority. One of the provisions of the law was that the Postmaster General could issue rules "as to the classes and description of advertisements which must not be broadcast." It was under this legislative authority that cigarette advertising was banned on television. In doing so the government argued that television advertising was particularly effective in encouraging adolescent smokers and that a ban was therefore an appropriate first step. It should be noted that commercial television was at the time far less important and well established in Britain than in the United States. Television was dominated by the state-owned and advertisement-free British Broadcasting Company (BBC); there was only one commercial station with a relatively limited number of viewing hours. In addition, there was a fair amount of anti-commercial, pro-BBC sentiment among the political elite in Britain.

The fact that the British first banned cigarette advertising, and later required warnings, while the United States did the reverse, is less important than the fact that the two countries dealt with these issues in very different ways. With the exception of the advertising ban, all subsequent British restrictions on advertising, promotions, and health warnings were accomplished through voluntary agreement between the government and the advertising or tobacco industries, rather than through formal regulations as in the United States. This is, of course, consistent with the approach that has been characteristic of most areas of economic policymaking in Britain. In the absence of similarly congenial relations, American health promotion policy, especially with regard to smoking and health, has been forced to rely on regulation.

PROMOTING HEALTH OR PROMOTING CIGARETTES?

The United States

HEALTH WARNINGS In 1965 Congress enacted the Cigarette Labeling and Advertising Act in an attempt to preempt a more restrictive and ambitious regulatory effort by the Federal Trade Commission (FTC). Shortly after release of the surgeon general's report, the FTC had scheduled hearings on regulations for the tobacco industry. The outcome of the hearings was a proposal to require a health warning on cigarette packages *and* in cigarette advertising. To prevent a much less amenable and accessible federal regulatory agency from determining the nature and scope of such warnings, the tobacco industry waged a successful campaign to move the decision out of the bureaucracy into the more accommodating halls of Congress. As described by Elizabeth Drew, what the tobacco industry set out to do was to graft "onto its built-in congressional strength from tobacco-producing states a sufficient number of congressmen to whom the issue was not one of health, or even of the tobacco industry, but one of curbing the powers of regulatory agencies, such as the FTC; . . . [and to throw] a heavy smoke screen around the health issue."[44] By successfully diverting attention from the health issue, the industry was able to defuse much of the opposition and passion that would later accompany other health promotion efforts.

The outcome of this strategy was that Congress preempted FTC action. It agreed to a health warning—about which the tobacco industry was not all that unhappy, since it might help in future liability cases—but one that was less foreboding and specific than those considered by the FTC. The warning would appear only on cigarette packages, not in advertising. Furthermore, the 1965 act prohibited the FTC, and state and local governments, from taking any action on labeling or advertising until 1969, when the law would expire. The stage was thus set for a repeat of the battle over warnings.

In 1969 the FTC again proposed more stern warnings for packages and the introduction of warnings on advertisements. Once again the Tobacco Institute engaged in a combined public relations and congressional lobbying effort belittling the link between smoking and health and challenging the legitimacy of administrative authority in this area. The results, however, were different this time. Initially it appeared that the tobacco industry was headed for another victory. The House of Representatives passed a bill that although requiring a somewhat strengthened warning on cigarette packages permanently prohibited states from banning cigarette advertising and delayed any regulatory agency action for six more years.

However, the industry ran into trouble in the Senate. According to Lee Fritschler, "The tobacco interests were regarded as having overplayed their hand [in the House victory]."[45] Perhaps more to the point is that the nature of the debate over smoking and health was changing. The tobacco industry continued to deny the relationship between smoking and ill health, but it was a link that few others now questioned. In the aftermath of the surgeon general's report the proportion of Americans smoking declined from 42 percent in 1965 to 38 percent in 1970, although it would again rise and peak in 1973. In addition, anti-smoking sentiment was evident elsewhere. In June 1969, at the time the House was considering the Public Health Cigarette Smoking Act, and six months before the Senate would, the California Senate approved a bill to ban cigarette advertising in newspapers and magazines, and on radio and television. Furthermore the *New York Times* had announced that it would no longer carry cigarette advertisements without health warnings.

[44] Elizabeth Drew, "The Quiet Victory of the Cigarette Lobby," *Atlantic Monthly* (September 1965): 76–77; see also A. Lee Fritschler, *Smoking and Politics*, 3d ed. (Englewood Cliffs, New Jersey: Prentice-Hall, 1983).

[45] Fritschler, *Smoking and Politics*, 140.

Thus, by the time Congress again considered labeling and advertising legislation, the political environment was quite different from that of 1965. Anti-smoking forces prevailed in the Senate debate over labeling and advertising, despite opposition from some of the Senate's most influential members—Sam Ervin (D-North Carolina), Harry Bryd (D-Virginia), James Eastland and John Stennis (D-Mississippi), Ernest Hollings (D-South Carolina), and Strom Thurmond (R-South Carolina). In December 1969, the Senate approved a bill that both strengthened the cigarette label warning and banned cigarette advertising on radio and television. Reconciliation of the two versions of the bill did not come until March 1970. The final bill strengthened the warning label and banned cigarette advertisements on radio and television. However, the new law also prohibited all state and local action on cigarette advertising and prohibited the FTC from taking any further action on printed advertising until July 1971. In addition, after that date the FTC was required to give Congress six months advance notice of any health-related regulations it was considering.

With passage of the Public Health Cigarette Smoking Act policy activity moved from the legislative to the administrative arena. In 1970 and 1971, following several skirmishes between the tobacco industry and the FTC, agreements were reached to include tar and nicotine content (December 1970) and a health warning (January 1972) in all cigarette advertising. For the most part Congress would not again be actively involved in the issue of labeling and advertising for more than a decade. (An exception was a 1973 congressional ban on little-cigar advertisements on radio and television.) Congress returned to the issue in the early 1980s when the FTC encouraged it to adopt more pointed and forceful warning labels. In a 1980 survey the FTC found that 47 percent of the women questioned did not know that smoking during pregnancy increased the risk of stillbirths and miscarriages, and 41 percent of the males surveyed were unaware that smoking shortens the life of the typical 30-year-old male.[46] The FTC and several anti-smoking groups urged that stronger health warnings replace the current label.

Between 1981 and 1984 a Congressional coalition of anti-smoking activists, including liberal Democrat Congressmen Henry Waxman of California and Edward Kennedy of Massachusetts, moderate Republican Senator Robert Packwood of Oregon, and conservative Republican Senator Orrin Hatch of Utah, whose Mormon background often finds him on the side of health advocates, introduced bills to require rotating health warning labels. These efforts were finally successful, despite Reagan administration opposition, with passage of the Comprehensive Smoking

[46] See *Wall Street Journal*, 11 May 1982.

Prevention Education Act of 1984, which required stronger, more specific, and periodically rotated warnings. Although the bill was held up in the Senate by North Carolina Senator Jesse Helms, it was finally approved after the Tobacco Institute quietly agreed to withdraw its opposition. Senator Hatch described the institute's move as displaying "great courage," but it was more likely a reflection of the fear that even more objectionable measures might follow later on if it did not capitulate.[47] The industry was particularly concerned about the imposition of higher cigarette taxes. In addition, as part of a compromise, anti-smoking forces agreed to support restraints on imported tobacco, which was threatening domestic tobacco farmers. The import restraint agreement was supported by cigarette manufacturers, who have increasingly been at odds with tobacco farmers.

ADVERTISING The issue of cigarette advertising lay dormant for even longer than the labeling issue. Following the 1970 ban on radio and television advertising, there were no significant efforts in Congress to expand the limitations on cigarette advertising beyond the electronic media. Cigarettes remained the most heavily advertised consumer product in the country. After 1971 tobacco companies merely shifted the money spent on advertising from radio and television to other forms, and ultimately vastly increased spending—from less than $300 million in 1970 to about $2.4 billion in 1988. Of this amount the largest single area of spending which currently accounts for over $1 billion, was promotional, for example, sponsorship of sporting and cultural activities.

Cigarette advertising and promotions promise to be a major battleground in the health and smoking debate in the 1990s. In January 1985 the National Advisory Council on Drug Abuse called for a total ban on cigarette advertising in newspapers, magazines, and billboards, and at concerts and sporting events. Over the next two years scores of health advocacy groups, including the AMA and the American Heart, Lung, and Cancer Societies echoed this demand. In 1986 two bills dealing with cigarette advertising were introduced. The first, and more limited effort, was introduced by Senator Bill Bradley (D-New Jersey) and Congressman Fortney Stark (D-California) and would have amended the Internal Revenue Code to disallow deductions for tobacco advertising expenses.

The second measure would have banned all advertising and promotion of cigarettes and other tobacco products. In June 1986 Congressman Mike Synar (D-Oklahoma) introduced the Health Protection Act, the first proposed total ban of cigarette promotions. The bill was given virtually no chance of passage, and at the time Congressman Synar admitted that

[47] *New York Times*, 27 September 1984.

"our intent is to continue to keep this issue at the forefront of public discussion while we search for the most effective means of discouraging tobacco use."[48] Despite the apparently symbolic nature of the effort, public hearings on the bill attracted large crowds of public health and tobacco and advertising industry representatives. Each side repeated familiar and conflicting claims about the nature of the health problem posed by smoking, the impact of advertising on consumption (as opposed to brand choice), and the rights of both smokers and nonsmokers. The tobacco and advertising interests in particular emphasized a link between the proposed ban and their First Amendment rights of freedom of speech and the press. This was in fact the first time that the issue of political or personal freedom arose in the debate on smoking and health. Up to this point the debate had been dominated by the credibility of the evidence linking smoking and ill health, and the economic consequences of the anti-smoking crusade on the tobacco and related industries. By interjecting the issue of both personal and corporate rights, the tobacco and advertising industries helped reformulate the debate to their advantage. That they were at least partly successful in this effort is reflected by the fact that in February 1987 the American Bar Association rejected a proposal to lend its support to a tobacco advertising ban.

Congressman Synar reintroduced his bill in 1989 with little success. The pressure to further restrict cigarette advertising and promotions will continue to be strong. In April 1987 two dozen members of Congress introduced a resolution urging the nation's broadcasters to resume the effective anti-smoking public service announcements that ran on television between 1967 and 1970. In April 1990 California began a multimillion dollar anti-smoking radio, television, and newspaper advertising campaign financed by a voter-approved twenty-five cent per package state cigarette tax. The campaign specifically targeted the same groups that the tobacco companies have identified as their most promising source of new smokers, young people, women, and minorities. Television spots have proven especially successful in the past, and the tobacco industry may well conclude that a ban is not such a terrible thing.

In November 1989 Senator Edward Kennedy introduced a bill (the Tobacco Products Control and Education Act) that would fund a $50 million advertising campaign to discourage smoking, as well as prohibit advertising aimed at young people. Although the evidence concerning the impact of advertising on consumption is ambiguous, policymakers are convinced that a relationship does exist, if only in terms of defining social acceptability, and it will remain one of the critical battlefields of the debate over smoking and health.

[48] *Congressional Record* (10 June 1986), 2020.

In early 1990 Dr. Louis Sullivan, the secretary of health and human services, initiated what promises to be a protracted and spirited debate when he urged the end to cigarette sponsorship of sports events. This issue, which has generated considerable debate over the years in Britain, has drawn relatively little attention thus far in this country. Critics charge that not only is the link between smoking and sports grotesque, but that sponsorship often has allowed the tobacco companies to circumvent the law prohibiting cigarette advertising on television. Cigarette brand names are currently prominently displayed during broadcasts of tennis, bowling and golf matches, baseball games, and automobile races.

Great Britain

ADVERTISING AND LABELING Initial British smoking and health policy took much the same path as U.S. policy. Within one month of the Royal College report *Smoking and Health*, the tobacco industry attempted to preempt government regulation, and improve its public image, by voluntarily agreeing to deglamorize cigarette advertising and eliminate it on commercial television before 9:00 P.M., when young people were most likely to be watching. As noted above, one of the few formal policy measures the British took was the 1965 Labor government ban on cigarette advertising in commercial television.

Labor governments during the 1960s—Labor was in power from October 1964 to June 1970—continued to focus on curbing promotional activities. In 1967 the government sought a voluntary agreement with the tobacco industry to end the use of gift coupons, which critics viewed simply as a reward for smoking more cigarettes. The health minister, in a tactic to be repeated many times in the next two decades, threatened legislation if a voluntary agreement could not be reached. The industry responded by suggesting that ten thousand jobs might be lost if promotional activities were ended. In a period when the British economy was already under considerable strain, the argument was quite compelling; coupons remained.[49]

Over the next two decades, under both Labor and Conservative governments, there would be a similar series of parries and thrusts, many of which ended in voluntary agreements and most of which left health advocates in a state of rage. According to Mike Daube, former director of Action on Smoking and Health, "The manufacturers for their part were fairly happy with a system which allowed them to negotiate agreements with successive governments, secure in the knowledge that negotiations could be dragged out for months and even years. The industry's main

[49] Quoted in Maurice Corina, *Trust in Tobacco: The Anglo-American Struggle for Power* (London: Michael Joseph, 1975), 33.

objective . . . was simply to avoid legislation."[50] In this regard the British tobacco industry has been completely successful. Except for the extension of the advertising ban to commercial radio in 1973, there has been no legislation since the 1965 ban on television advertising.

The most persistent issues in the long-running debate have involved limitations on where cigarettes could be advertised, how much the industry should spend on advertising, what, if any, promotional activities would be allowed, the appropriateness of cigarette sponsorship of athletic and cultural events, and the wording, location, and size of health warnings. Health warnings evolved over time in terms of their specificity, ferocity, and variety. The first warnings appeared after a 1971 voluntary agreement and over the next fifteen years were changed in form, content, and location, and extended to other types of advertising. (The most recent change was an increase in the size of warning labels on cigarette packages, and it was imposed by the EEC. It not only earned a public rebuke by Thatcher, who has become an increasing critic of the EEC, but may well represent a new and promising avenue for health advocacy in Britain and elsewhere in the Community.)

Promotional sponsorship of sports, and to a lesser extent cultural events, has been of particular concern to anti-smoking forces because, they claim, it has been used by the cigarette industry to circumvent the television advertising ban. Sponsorship of televised sporting events keeps the names of cigarette brands on television for hundreds of hours per week. Health promotion activists argue that advertising should be banned except, perhaps, at the point of sale, that all promotional activities, including gift coupons and contests, should be prohibited, and that there should be no sponsorship of either cultural or sporting events by cigarette companies. This position was endorsed in the 1976 Commons Expenditure Committee report on preventive medicine but was not adopted by the Labor government.

In May 1979 the Conservatives came to power and thus became responsible for negotiating the next voluntary agreement or adopting legislation. By now pressure for legislation was strong. In November 1979, at the time when the government started talks over a new agreement, a Labor M.P. introduced a private member's bill that would have given the health minister broad powers to regulate or ban tobacco advertising and sponsorship. The idea of a total ban had the support of the Royal College of Physicians, the BMA, and an all-party Parliamentary Group on Smoking and Health. In addition, there was some indication of support within the Thatcher government itself. Earlier in the year the under-secretary for

[50] Mike Daube, "The Politics of Smoking: Thoughts on the Labour Record," *Community Medicine* 1 (November 1979): 306.

health, Sir George Young, spoke before a World Health Organization conference on smoking and health in Stockholm. In that speech Sir George said, "There is a very clear indication that the advertising of tobacco products has a specific effect, not necessarily on confirmed smokers of 20 or 30 years but particularly in getting young people to start smoking." He went on to say that "the tobacco industry knows full well that intransigence on its part could be counter-productive. We are determined to make progress on advertising."[51]

Talks between the government and the tobacco industry began in November 1979. One year later an agreement had still not been reached. Pressure from opposition members within Parliament and public health interest groups remained intense throughout the period, but it became increasingly clear that the government was opposed to legislation and reluctant to impose economically crippling restrictions on a successful and important revenue-producing industry.

Government representatives and tobacco manufacturers finally reached an agreement in November 1980. The agreement was described by Health Ministry officials as a "modest advance on the road to reduce the toll of disease and death from smoking while leaving the House free to take further steps in the same direction as may be appropriate."[52] The new agreement, which was to last until July 1982, included a commitment to cut spending on cigarette poster advertising to 70 percent of the 1979–1980 level; introduction of new, rotating health warnings for packages and posters, and 50 percent more room on the posters for the warnings; an end to advertising for high-tar cigarettes; and an industry agreement to keep advertising posters away from schools and playgrounds. Finally, the tobacco companies agreed not to advertise on television other tobacco products such as cigars and pipes that bore the same name as their cigarette brands. The latter was a practice that anti-smoking groups deplored as a clear circumvention of the ban on cigarette advertising. The agreement did not, however, impose any restrictions on sponsorship. The overall reaction of health groups to the new agreement combined rebuke and frustration. The general consensus among observers was that once this agreement lapsed it would almost certainly be followed by legislation to ban, or at least severely limit, advertising and sponsorship.

That such was not to be became evident in two episodes following the new agreement. The first was the defeat, by Conservative backbenchers in June 1981, of a Labor M.P.'s private member's bill that would have given the government the power to control tobacco advertising and spon-

[51] Sir George Young, "The Politics of Smoking," *The Smoking Epidemic*, ed. Lars Ramstrom (Stockholm: Almquist and Wiksell International, 1980), 125.

[52] *Times* (London), 22 November 1980.

sorship. The bill's sponsor, Laurie Pavitt, thought his bill had a reasonable chance of success despite the obstacles facing private member's bills. Pavitt's optimism was based in part on indications from Patrick Jenkin, the social services secretary, that the government would allow a free vote on a bill to curb advertising. Pavitt believed that there was sufficient anti-cigarette sentiment among backbenchers in both parties in the House to win. However, on the day his bill was to receive a second reading, he discovered that the bill proceeding his, a Zoo Licensing Bill, had attracted 164 proposed amendments. Debate over the Zoo Bill, and all its amendments, left no time for consideration of the Tobacco Advertising Bill. One of the sponsors of the amendments, an adviser to the British American Tobacco Company, defended his action by saying that he had "always been interested in animal welfare."[53]

The second of the two actions occurred eleven months after the new agreement, when Jenkin and Sir George Young were replaced in the Department of Health and Social Security by men who were much less likely to take a hard-line position with the cigarette companies or to support legislation. Although Thatcher and others in her government denied that the Cabinet shuffle had anything to do with the position of the health officials toward smoking, their assertion was questioned by many in the media. Whether or not Jenkin and Young were replaced for their apparent anti-smoking sympathies is unclear. It is certainly the case that the new health minister, Dr. Gerald Vaughan, whose constituency includes a large brewery owned by the Imperial Group, Ltd. (formerly Imperial Tobacco), one of Britain's largest cigarette manufacturers, made it quickly known at the annual Conservative Medical Association meeting that he had no intention of seeking legislation to curb advertising and that he would instead seek another voluntary agreement with the tobacco industry.[54]

By the early 1980s the issue of sports sponsorship had become particularly contentious. Sponsorship of televised sporting events gave the cigarette companies a several-fold return on their investment and access to a medium from which they were theoretically banned. Of particular concern to the critics of sports sponsorship was its impact on smoking among young people. Studies showed that school children had a higher rate of smoking (41 percent) than either adult males (36 percent) or females (32 percent). Furthermore, a high proportion of these children, between one-third and one-half, watched cigarette-sponsored sporting events. Anti-smoking groups complained that the main purpose of such sponsorship was to woo young people into the smoking habit.

[53] *Times* (London), 13 June 1981.

[54] *Manchester Guardian*, 16 November 1981; *The Observer* (London), 11 November 1981; *Times* (London), 28 February 1984.

In March and October 1982 the Thatcher government concluded two new voluntary agreements relating to tobacco sponsorship of sporting events and advertising in general. The first provided for an increase in the amount of money the tobacco companies could spend on prizes in sports sponsorship. The increase, it was argued, was an adjustment for inflation, and the real value of the money would remain constant. The one concession won by the government was that all advertisements for tobacco-sponsored events would have to carry a health warning. The second agreement covered direct cigarette advertising. The main provisions of the agreement included a commitment not to use video cassettes or satellite broadcasting for advertising, and a progressive reduction on expenditures for poster and cinema advertising. In addition, the tobacco industry agreed to establish a trust fund of £11 million ($19.3 million) over the duration of the agreement, for research on health promotion. Incredibly enough, the fund could not be used for studying the effects of tobacco on health.

As the time approached for yet another round of negotiations over the next voluntary agreement, public opinion polls indicated increased support for dropping sports sponsorship. In May 1986 the London *Times* commissioned a poll that found that 33 percent of those interviewed favored a total ban on tobacco sponsorship of all sporting events, and an additional 11 percent favored a ban on sponsorship of televised events. The total of 44 percent favoring a television ban was up from 24 percent in 1975 and 34 percent in 1981.[55] The minister for sport indicated that if the tobacco industry and broadcasters could not better control abuses of obvious advertising, the government would call for legislation.

If health promotion advocates were encouraged by this attitude, they were to be disappointed. In January 1987 the government announced a new voluntary agreement that would run at least until October 1989. The main provisions of the agreement included a cut in expenditures on sponsorship, tighter controls over the location of brand-name signs in televised events, and a 50 percent increase in the size of health warnings on the signs. In addition, tobacco companies could not sponsor events designed to appeal specifically to persons under 18 years of age, and media advertising of sponsored sporting events could not depict any participants actually engaging in a sport. Finally, the new agreement would be monitored by a government-appointed Committee for Monitoring Agreements on Tobacco Advertising and Sponsorship. The agreement addressed some of the major concerns of opponents to sponsorship, but most still believed that only a total ban would disassociate sports and smoking, especially in

[55] *Times* (London), 31 May 1986.

the minds of young people. The sponsorship agreement was attacked by anti-smoking groups as a "major victory for the tobacco industry."[56]

The sponsorship agreement was tied to an earlier voluntary agreement on advertising, which had been reached in April 1986. It addressed two increasingly sensitive areas, namely smoking among young people and women. The agreement contained the following provisions: (1) during the agreement, the industry would spend £1 million per year ($1.5 million) on a campaign to enforce the decades-old, but rarely followed, prohibition of cigarette sales to children under sixteen years of age; (2) there would be a stronger enforcement of the ban on posters near schools; (3) the industry would stop using cigarette brand advertising on logos or "give-aways" to children at events sponsored by tobacco companies; (4) there would be no cigarette advertising in any magazine for which one-third or more of the readers were between the ages of fifteen and twenty-four years old; and (5) there similarly would be no cigarette advertising in magazines with a female readership of over 200,000.

Much like the sponsorship agreement, this one would be monitored by a new committee consisting of government and tobacco industry representatives, chaired by an independent chairperson. The committee would receive complaints about implementation of the agreement and conduct periodic surveys of its observance. Clearly the government was anxious to enhance the integrity and credibility of the voluntary agreement approach. In March 1988 the Committee for Monitoring Agreements on Tobacco Advertising and Sponsorship issued a report alleging over 460 violations of the voluntary agreement. In addition, an earlier study found that several women's magazines were not in compliance with the agreement. The study concluded that "the voluntary agreement isn't working. It is not worth the paper it is written on."[57]

In the three decades after the first widely publicized evidence linking smoking and ill health, both American and British governments alerted consumers to the health dangers posed by smoking, reduced the temptation to smoke by removing cigarettes from the more seductive advertising media, and tried to deglamorize the product by disassociating it from sporting and other events (see Table 4-1). The latter effort has been the least successful thus far. There have been two major differences between the two countries. The first has been in the political vehicles used to accomplish these goals: regulation and voluntary agreements. The second is that the British have gone further in trying to remove cigarette advertising from the public's view.

[56] Ibid., 20 January 1987.

[57] Ibid., 24 March 1988 and 7 October 1986.

TABLE 4-1
Major Changes in Cigarette Labeling and Advertising Policy, United States and Britain, 1965–1987

		United States		*Britain*
Labeling changes	1965	Warnings required on cigarette packages	1971	Warnings required on cigarette packages
	1984	Rotating warnings required	1980	Rotating warnings required
Advertising bans	1970	On radio and television	1965	On commercial television
			1967	In newspapers and magazines
			1973	On commercial radio
			1980	All advertising of high-tar cigarettes
			1986	In magazines with primarily young adult or female readers
			1987	Sponsorship of events designed to appeal to people under eighteen years of age

What have been the consequences of these differing approaches? In the area of labeling, advertising, and promotions it appears voluntarism has better served the interests of the tobacco industry than the public's health. First, and contrary to Vogel's conclusion that voluntary agreements have been as successful in Britain as regulation has been in the United States, many observers have been skeptical of the success of self-regulation in the case of the tobacco industry. Michael Calnan, of the University of Kent, has argued, "Not only are there doubts about whether the voluntary agreements are operating effectively, but some authors have suggested that the agreements have some benefits for the tobacco industry."[58] As I have shown, the tobacco and advertising industries have been remarkably successful in continuing to promote tobacco products, especially through cultural and sports sponsorship, despite apparent agreements not to do so.

Furthermore, although voluntary agreements appear attractive because they suggest a less conflictual relationship between government and industry, they have probably produced more conflict than regulation has in the United States. The reason is that most of the voluntary agreements are negotiated for finite periods of time. Thus these issues periodically reappear on the policy agenda, rekindling the conflict every two to four years as British governments, the tobacco and advertising industries, health ex-

[58] Michael Calnan, "The Politics of Health: The Case of Smoking Control," *Journal of Social Policy* 13 (July 1984): 281. See also Peter Taylor, *Smoke Ring*.

perts, and anti-smoking groups battle over acceptable limitations on the promotion of tobacco products.

In the United States, anti-smoking measures have relied exclusively on either legislative or administrative regulation. This approach has instigated some self-regulation on the part of the tobacco industry, but for the most part changes in the sales and promotion of cigarettes have been the result of formal government activity. Although certain issues, such as warning labels on cigarette packages, have reappeared, the general issue has not occupied as much attention in the United States as in Britain. Nevertheless, the British print media is relatively freer of cigarette advertising as a result of voluntary agreements. Publishing interests in the United States are highly addicted to cigarette advertisements, and even the most socially conscious, like *Ms* magazine, have been unwilling to support a ban. The success of the U.S. cigarette and advertising industries has been largely due to their ability to couch this issue in the idiom of free speech, one of the nation's most cherished political values.

Taxation: "The Goose That Lays the Golden Egg"

Taxes have long played an important part in the politics of smoking. The relationship among price, taxation and consumption has been well documented. Reviewing the U.S. literature, Kenneth Warner concluded that "a wealth of empirical research indicates that the number of cigarettes smoked varies inversely with changes in the price of cigarettes, and tax variations constitute a major component of price differences."[59] However, if higher cigarette taxes discourage consumption, they also ultimately reduce overall government revenues. In both Britain and the United States policymakers have been warned about "killing the goose that lays the golden egg."[60]

CIGARETTE TAX POLICY IN THE UNITED STATES

Historically tobacco taxes in the United States have been viewed mainly in terms of revenue raising, rather than enhancement of morality or health. Although there had been a temporary federal tax on tobacco as early as 1794, it was not until during the Civil War (1863) that the federal government imposed what was to become a permanent tobacco tax. Initially tobacco taxes provided a substantial proportion of total internal

[59] See Warner, "Smoking and Health Implications," 1028. Also see Calnan, "The Politics of Health," 292–93; and Great Britain, Confidential Report by the Interdepartmental Study, "Cigarette Smoking and Health" (1971): 6.

[60] See, for example, comments by Congressman McConnell (R-Ky) in *Congressional Record* (19 September 1986), 13002.

revenues. In the 1870s and 1880s they accounted for nearly one-third of all federal internal revenue collections. Over the years, and especially after the introduction of the income tax, revenue from tobacco taxes as a proportion of total federal revenues has decreased substantially, from 19 percent in 1915 to just .4 percent in 1987. Despite their relative decline, tobacco taxes, which raised $4.7 billion in 1988, are seen as one part of the "smoke ring" inhibiting governments from going too far for fear of ultimately reducing total revenues.

The decline in the relative importance of tobacco taxes has not immunized them from the attention of anti-smoking advocates. In 1983, for the first time in nearly three decades, the federal excise tax on cigarettes was increased from 8 cents to 16 cents a package. The increase was to be a temporary two-year measure to help offset the enormous deficits created during the Reagan administration. The deficit problem did not get any better, and the Reagan administration and tobacco-state legislators reluctantly agreed to make the increase permanent. Revenue "enhancement," however, was only one issue behind support for the increase. Senator Mitch McConnell (R-Kentucky), who opposed still another proposed increase in 1986, complained to his colleagues that "the plain truth of the matter is that the proponents of the cigarette tax increase are again dragging up the health issue and are covering it up with the smoke of the budget deficit."[61]

Tobacco taxes have been a concern of state-level health and fiscal policymakers as well. In Iowa, which in 1921 became the first state to adopt a tobacco tax, and the ten other states that quickly adopted a tobacco tax, "cigarette taxation was viewed as a last desperate stand against the 'cigarette fiend.' "[62] Later on morality gave way to economic necessity as the states, especially in the 1930s and 1940s, sought ways to provide property-tax relief. Over the years tobacco taxes became an important revenue source for most of the fifty states and a frequent choice for shoring up shaky revenue bases. In fact, tobacco taxes became relatively more important to the states than to the federal government. In 1945 the federal government collected $932 million in tobacco taxes and the states $145 million. By 1975, however, with no increase in the federal tax since 1951, the states were collecting more ($3.4 billion) than the federal government ($2.3 billion). The gap between federal and state revenues again declined following the doubling of the federal excise tax in 1983. Despite the absolute increases, state income from tobacco taxes has declined from 2.8 percent in 1960 to .9 percent in 1985. Finally, it should be noted that both the rate and absolute amount of federal and state cigarette taxes

[61] Ibid., 13001.

[62] Quoted in Kenneth M. Friedman, *Public Policy and the Smoking-Health Controversy* (Toronto: Lexington Books, 1975), 17.

have increased over the years, but the proportion of the retail price of a package of cigarettes accounted for by taxes has declined from about 50 percent in the 1950s and 1960s to about 31 percent in 1986. This declining rate of taxation, which Larry White attributes to the "political might of the tobacco lobby," has also made cigarettes one of the nation's most profitable products.[63]

In the 1980s health considerations began to play a more prominent role in cigarette tax policy. In 1987, for example, thirty-one states considered cigarette excise tax increases, almost all of which were promoted as efforts to discourage consumption, and eleven states adopted them. In November 1988 California voters approved a twenty-five cent per pack increase in the state cigarette tax, despite a $20 million effort by the tobacco lobby to defeat the ballot measure. The increase brought the California cigarette tax to thirty-five cents per package, making it one of the highest in the nation. It should be noted that the national range is quite broad with a high, as of July 1, 1989, of forty cents per package in Connecticut and a low of two cents in North Carolina.[64] Over the next few years one can expect all levels of government to raise cigarette taxes both to balance their budgets and to promote health. It has been estimated, for example, that a seventy-five cent per pack increase in the federal excise tax would produce an additional $13.3 billion in revenues and a 40 percent decline in consumption.[65] And unlike advertising bans or "harassment of smokers," tax increases are generally acceptable among conservatives.

CIGARETTE TAX POLICY IN GREAT BRITAIN

In Great Britain, too, cigarette taxes have played an important role in the debate on smoking and health. In his 1979 speech before the Stockholm conference, Sir George Young spoke of various measures to discourage cigarette smoking. The junior health minister argued that "perhaps the biggest weapon that a Government has at its disposal is the ability to fix the price at which cigarettes are sold."[66] The public pressure to use that weapon has been particularly intense in Britain over the last two decades. But almost as intense has been pressure from within government not to kill the goose that lays the golden egg.

As noted earlier, the interdepartmental committee studying cigarette smoking and health in 1971 chose not to take steps to discourage smoking for fear of the loss in tax revenues. Their report concluded that although the material and human costs of smoking were high, a successful anti-smoking campaign would be economically more costly in the long

[63] White, *Merchants of Death*, 195.
[64] *New York Times*, 11 April 1990.
[65] See *Fortune* (27 February 1989).
[66] Young, "The Politics of Smoking," 123.

run in terms of an increase in health care and social security costs resulting from a larger and more elderly population. In addition, the report pointed out that even a modest price increase of five pence per package would add one percent to the Retail Price Index. The Retail Price Index is the yardstick of inflation and the basis for social security and cost-of-living increases for government and some private employees. As it goes up, so do public expenditures and overall inflation.

In an adjournment debate on May 16, 1980, a Labor backbencher, Tristan Garel-Jones, raised the issue of the relationship between the RPI and government efforts to discourage tobacco and alcohol consumption. Garel-Jones noted that it was an "open secret" that the chancellor of the exchequer "wished to place heavier duties on tobacco and alcohol in his recent Budget but was inhibited from doing so by the effect it would have on the Retail Price Index."[67] He further noted that because previous chancellors suffered similar inhibitions, the relative cost of tobacco and alcohol had declined over the years. Much of the debate focused on the feasibility of removing tobacco and alcohol from the RPI so that tax increases would not affect the index. Ultimately the minister of state for the Treasury accepted the principle that tobacco consumption should be discouraged but went on to remark, "We must take into account certain realities." One of those realities was that tobacco would remain part of the retail price index.

In the United States increases in the federal tax on cigarettes have been few and far between, but in Britain the issue has become an annual one. Since the early 1970s each new budget has rekindled the debate over balancing fiscal needs and health concerns. In virtually every year since the early 1970s the duty on cigarettes has increased; the 1987 election-year budget was a notable exception. Critics argue, however, that the increases have not been sufficient to discourage consumption. The general assumption is that in order to do so the increase must exceed that of overall price increases. Table 4-2 compares average annual tobacco price increases—almost all of which can be accounted for by tax increases—and overall price increases. As indicated in the table, in the decade from 1961–71, the price of cigarettes rose annually by an average of 3.9 percent, while the overall rate of inflation was 4.7 percent. From 1971 to 1980, average annual tobacco price increases continued to lag behind overall inflation, 13.0 compared to 14.1 percent. It was only beginning in 1980 that the absolute price increase (e.g., seventeen pence, or thirty-four cents), and the comparative price increase (23.5 percent compared to 11.9 percent overall inflation) was to have the dramatic impact on cigarette sales for

[67] *United Kingdom Parliamentary Debates* (Commons) (Tobacco and Alcohol) (16 May 1980), col. 1974.

TABLE 4-2
Index of Retail Prices, Average Annual Rates of Change, Overall and for Tobacco, Great Britain, 1961-1984 (in Percent)

	1961-1965	*1966-1971*	*1971-1976*	*1976-1980*	*1980-1981*	*1981-1982*	*1982-1984*
All items	3.6	5.7	14.5	13.8	11.9	8.6	4.8
Tobacco	4.9	2.8	12.0	14.1	23.5	15.4	10.9

Source: Great Britain, Central Statistical Office, *Social Trends 16* (London: Her Majesty's Stationery Office, 1986), 95.

which health promotion advocates had been calling: sales declined 15 percent from 1981 to 1982.

The dramatic seventeen-pence increase in cigarette taxes in 1981 was followed by more modest increases (five pence in 1982 and three pence in 1983), reflecting the reality that only so much of an increase is politically healthy. The relationship between tax increases and politics was revealed in 1987, when in anticipation of a general election the Chancellor of the Exchequer chose not to increase the cigarette (or whiskey) tax at all. This decision led the conservative *Economist* to describe the new budget as "nice for vice" and antismoking groups to express their "disappointment."[68]

Despite a variety of idiosyncratic factors affecting cigarette tax policy, the overall trend has been toward increasing taxes, at least in part for health promotion purposes: between 1979 and 1987 the real value of duties on cigarettes increased by 50 percent. But perhaps more important from a comparative perspective is that in Great Britain cigarette taxes account for over 75 percent of the price of a package of cigarettes, compared to about 35 percent in the United States. In addition, although revenue considerations are by no means inconsequential, in Britain most of the increases in recent years have been prompted, at least in part, by health concerns. In his 1986 budget the Conservative Chancellor of the Exchequer, Nigel Lawson, made it clear that the eleven-pence increase in the cigarette tax was "for health reasons."[69] Anti-smoking forces have been dissatisfied with the speed and magnitude of these increases, but tax policy recently has become one of the main weapons in British health promotion efforts.

Political Issues: Where Do You Stop?

The debate on smoking and health has been dominated by economic issues. However, it has never been exclusively a matter of public health

[68] *Economist*, 21 March 1987, 54; *Times* (London), 18 March 1987.
[69] *Times* (London), 19 March 1986.

versus government or corporate wealth, and it is likely to be even less so in the future. There are several interrelated reasons for the decline, but not disappearance, of the economic dimension of the issue. To begin, cigarette consumption is down in both countries, and it is expected to continue to decline. It is estimated that in Britain consumption in 1994 will be about 66,000 million cigarettes, down from 132,000 million in 1975. It seems highly unlikely that either country will be "smoke free" by the year 2000, but it is clear that smoking will become the habit of a minority, perhaps 10 to 15 percent of the population.

The drop in consumption has also led to a decline in the number of people dependent upon tobacco for their livelihoods. In Britain, direct employment in cigarette manufacturing has declined from about 41,000 jobs in 1974 to about 21,500 in 1986. There have been similar changes in the United States. In North Carolina, the nation's largest tobacco producing state, farmers are increasingly turning to other crops: in 1988 tobacco accounted for just 19 percent of total farm income in the state, down from about 50 percent in the mid-1950s. Recently the North Carolina State Goals and Policy Board called tobacco a "crop whose future is in jeopardy." The Board noted ominously that "we must grasp the nettle and face the very real possibility that tobacco production as a major enterprise may end in North Carolina and that many tobacco growers may be forced either to leave the land or adapt to other agricultural income sources."[70] It is important to remember, however, that there always have been significant noneconomic dimensions to the debate over health and smoking, and although anti-smoking advocates often dismiss them as mere camouflage for the economic issues, it is a mistake to do so. Cigarette smoking, despite the large economic shadow that it casts on the political landscape, is still essentially a life-style issue, and as such activates many of the same legitimate political concerns as other health promotion issues.

The libertarian argument that individuals should be free, even to be foolish, without state interference in their lives, has often been accompanied by warnings of the dangers of the slippery slope of social control. The following are representative expressions of this concern from a parliamentary debate on smoking and health:

> Where would we stop in trying to interfere with products, which, if used to excess, can be dangerous? Alcohol, of course is an obvious example, but there are plenty of others that are less obvious—man-made fibres, aerosol sprays, butter, milk, fish and chips. All those things show statistical links with one serious illness or another. I do not want us to drift down a slippery, Scandinavian-style slope.[71]

[70] *New York Times*, 5 April 1987.

[71] *United Kingdom Parliamentary Debates* (Commons), vol. 984 (9 May 1980), col. 758.

> If we are to take individual items that affect the health of the nation, why do we single out tobacco? Why do we not move on, quite naturally, to the next item, which is alcohol? How many people die from alcoholism? Once one starts, where does one stop?[72]

> There are many other areas . . . where Parliament could legislate if once we went down that road and interfered with these freedoms.[73]

U.S. legislators too worried about "where it would all end." In opposing a bill that would have required cigarette manufacturers to list the chemical additives in cigarettes, expand warnings, and make permanent the status of the Office on Smoking and Health, Democratic Senator Wendell Ford of Kentucky warned his colleagues that "this legislation would establish the Secretary of HHS [Health and Human Services] as a czar of personal behavior to co-ordinate a multitude of government programs, research projects, and behavioral studies to tell people what's good for them. Smoking happens to be today's targeted behavior, but who knows what tomorrow may bring?"[74]

It is easy to dismiss many of these arguments ad absurdum as ideological dressing for pecuniary interests, but it would be wrong to do so in all instances. There is genuine concern among conservatives in both countries that the social engineering implications of cigarette legislation are an integral part of an overall tendency to allow the state to accomplish what undisciplined individuals can or will not. And it is difficult to refute these claims, for the regulation of cigarettes is part of a broader health promotion movement that *is* highly paternalistic in its assumptions and proposed policies. Whether or not one accepts these paternalistic policies as necessary for the collective good is a function of one's ideological position.

Smoking in Public Places: Whose Freedom Is It?

Restricting the use of cigarettes in public places for health reasons is the most recent, contentious, and politicized issue in the debate over smoking and health. This issue, more than advertising, health warnings, or taxation, captures the essence of the politics of health promotion because it directly impinges on personal freedom: the freedom to smoke, and the freedom to be free of the smoke of others. This "second generation" issue has been far more prominent in the United States than in Britain for two reasons. First, as mentioned in Chapter 1, the level of health conscious-

[72] Ibid., col. 722.
[73] Ibid., col. 745.
[74] Quoted in Taylor, *Smoke Ring*, 237.

ness and the life-style movement are simply stronger in the United States. The health ethic, which has its roots in the country's periodic (and current) return to moral puritanism and substance prohibition, has produced more aggressive policy proposals. Second, the decentralization of political power in the United States has facilitated the aggressive anti-smoking campaign by allowing health advocates to bypass the congressional tobacco lobby and pursue their agenda at the state and local level, an option not available in Britain.

THE UNITED STATES

For the tobacco industry the attempt to limit the use of cigarettes is the most serious threat since publication of the surgeon general's and Royal College's first reports. Aside from the practical implications of restrictions on smoking in public places, these limitations reflect and fuel the transformation of cigarette smoking from a glamorous and socially acceptable practice to a scorned one. In addition, this issue raises the question of the negative externalities of smoking, including both social and personal costs and harms, thus threatening to transform the entire nature of the debate.

The growing social unacceptability of smoking has been prompted by what many see as increasingly credible claims about passive (alternately called environmental, secondary, or involuntary) smoking. This issue has transformed cigarette smoking from a putatively self-regarding to a demonstrably other-regarding activity with significant negative externalities. As a result, the tobacco industry has devoted a considerable amount of its energies and resources in recent years to counter the claims concerning "environmental tobacco smoke." In November 1985 R. J. Reynolds Tobacco Company placed full-page advertisements in major newspapers around the United States. Entitled "Passive Smoking: An Active Controversy," they reviewed the results of two scientific meetings dealing with passive smoking. The advertisement quoted from the summary of one study that found that "the connection between [environmental tobacco smoke] and lung cancer has not been scientifically established to date." The Reynolds advertisement then expressed the concern that is at the heart of this issue: "For today, many non-smokers who once saw cigarette smoke merely as an annoyance now view it as a threat to their health. Their growing alarm is being translated into heightened social strife and unfair anti-smoker legislation. We believe these actions are unwarranted by the scientific facts—and that it is rhetoric, more than research, which makes passive smoking an active controversy."[75]

[75] *New York Times*, 13 November 1985.

Unfortunately for the tobacco industry, this is not an opinion shared by the U.S. surgeon general. In December 1986, the surgeon general issued a report on the health consequences of involuntary smoking and concluded that it was more than merely an "annoyance." It "is a cause of disease, including lung cancer, in healthy nonsmokers," and of increased frequency of respiratory infections among children of smokers. The surgeon general declared that "the right of a smoker to smoke stops at the point where his or her smoking increases the disease risk in those occupying the same environment."[76] There continues to be controversy surrounding passive smoking, but the impact of the issue, especially in the United States, has been considerable. That the tobacco industry takes this particular challenge seriously is reflected in their response to the surgeon general's report on involuntary smoking. A spokesman for the Tobacco Institute responded to the study by accusing Surgeon General Koop of abusing "his position, placing his political agenda above scientific integrity. No previous Surgeon General has exploited his office in such a blatantly propagandistic manner."[77]

Public policies limiting smoking in public places began earlier and have moved more rapidly and extensively in the United States than in Britain. The first significant effort in this area predated the surgeon general's report and occurred at the state and local level, especially where tobacco industry influence was weak. In 1975 Minnesota passed the nation's first Clean Indoor Air Act, which prohibited smoking in all public places except for specially designated areas. A number of other states and localities adopted limited smoking bans, in elevators, public transportation vehicles, or libraries, for example, but policy activity in this area remained tentative through the late 1970s and early 1980s. In the mid-1980s federal, state and local governments began a rush to legislate. It was at this time that the evidence implicating passive or involuntary smoking became more credible, and the image of smoking more tarnished. Starting in the mid-1980s many states and municipalities began to regulate smoking in various public settings, including government buildings, restaurants, public transportation, health-care facilities, and the workplace. By 1990 hundreds of cities and counties and nearly all of the states had imposed limitations on smoking in public places. Among these was the state of Virginia, the cradle of the tobacco industry, which in March 1990 enacted a law limiting smoking in a variety of public settings. In Utah, Minnesota, Montana, and Nebraska, smoking is prohibited virtually everywhere except outdoors and in private homes.

In response to this most recent and threatening assault on smoking, the

[76] Ibid., 17 December 1986.

[77] Ibid., 4 January 1987.

tobacco industry has launched a campaign to redefine the entire smoking issue into one of personal (and corporate) freedom. In 1988 the Tobacco Institute began a nationwide "Enough is Enough" campaign emphasizing the oppressed status of the American smoker. The newspaper version of this campaign listed six "enoughs": taxation, censorship, legislation, harassment, control, and discrimination. The theme, along with record spending by tobacco manufacturers on a state ballot measure, proved successful in Oregon, where in October 1988 voters defeated a measure that would have imposed a ban on virtually all indoor smoking. The measure, despite strong initial support in the polls (65 percent in June 1988) was decisively defeated by a 63 to 47 percent margin. Analysts credit the campaign's emphasis on personal freedom for the measure's defeat.

In the fall of 1989 the tobacco industry lost a major battle on the passive smoking front when Congress approved a ban on smoking on all domestic airline flights. The ban, which went into effect in February 1990, replaced a more limited one that had been approved in 1987 and applied to flights of two hours or less. The ban was one of the most significant defeats in years for the tobacco lobby at the national level, and is a measure of just how vulnerable the industry has become.

Finally, it is a measure of how far the nation has traveled since 1964, and how less influential the tobacco lobby has become, that even the federal government during the Reagan administration involved itself in the campaign to limit smoking. Surgeon General Koop called for a smoke-free society by the year 2000, and although his aggressiveness caused considerable discomfort in the Reagan White House, his position clearly had an impact on federal policy. In December 1986 the General Services Administration, which had been considering a smoking ban in Federal buildings since 1981, announced regulations requiring heads of agencies occupying sixty-eight hundred federally owned or leased buildings, where nearly one million government employees worked, to provide a "reasonably smoke-free environment" for federal workers. Each agency was given freedom to determine how best to achieve this goal, but the general rule was to prohibit smoking in all but designated areas.

However, it was not the General Services Administration but the U.S. Army that took the first significant steps toward a smoke-free environment for federal employees. In June 1986, the secretary of the army announced that in compliance with the secretary of defense's "health promotion program" the army would make "nonsmoking the norm." The secretary ordered that smoking would either be prohibited completely or limited to designated areas in military offices, barracks, vehicles, and aircraft. The directive applied to both military and civilian employees of the army and was justified because "smoking tobacco harms readiness by impairing physical fitness and by increasing illness, absenteeism, premature death and health care costs."[78] Apparently the army was concerned about the negative externalities of smoking and their implications for both national defense and budget balancing.

GREAT BRITAIN

As noted in Chapter 1, a long-standing issue among students of comparative politics and public policy has been whether the structural differences between unitary and federal systems produce different policy results. In the case of smoking and health, the answer appears to be that they do. The British have lagged behind the United States in imposing restrictions on smoking in public places, in part because there has been less innovation and autonomy at the local level. In both countries tobacco interests have exercised their most successful efforts at the national level. In both cases this message has been carried by national legislators from either tobacco-producing areas, in the case of the United States, or manufacturing areas, in both Britain and the United States. In the United States, however, anti-smoking groups have been able to break the grip of the tobacco industry in those states and localities where tobacco-producing and -manufacturing interests are weak or nonexistent. It is not at all surprising then that the earliest or most aggressive restrictions on smoking in public places came in states like Minnesota, Nebraska, and North Dakota, and cities like Chicago, Los Angeles, and San Diego. These early efforts were followed by more aggressive laws in Montana, Oregon, Connecticut, Maine, Iowa, New Jersey, Utah, Alaska, Hawaii, and New York—not a tobacco state in the lot.

In Britain, on the other hand, there are very few restrictions on public smoking, and these mainly in areas where either safety or potential social nuisance factors are particularly compelling. Beginning with a British Rail ban on smoking in restaurant and buffet cars in 1982, much of the British public transportation industry, including ferries, airlines, buses, and intracity trains have imposed restrictions, typically in the form of separate smoking and nonsmoking areas. In 1984, following a costly fire at one of its stations, the London Underground imposed a complete ban on smoking in its cars and all underground railway stations. Despite these efforts, and some restrictions in entertainment facilities, smokers face far fewer restrictions in Britain than in the United States, where policy innovation has percolated upward. In Britain, where it will take parliamentary action to accomplish widespread restrictions, the tobacco lobby has been able thus far to defeat the few backbench efforts at legislating a ban on smok-

[78] See ibid., 12 June 1986, 13 June 1986, and 22 June 1986.

ing in public places.[79] In the absence of the aggressive moralizing characteristic of the anti-smoking crusade in the United States, there has not yet been sufficient sentiment to overcome structural barriers to a more restrictive British policy. Recently, however, the tide of scientific and lay opinion has begun to change. Following a report by the government-sponsored Independent Scientific Committee on Smoking and Health, which endorsed the position that passive smoking poses a public health risk, the BMA called for a workplace ban on smoking. In addition, members of Parliament promised to pursue a private member's bill to ban smoking in various public settings. In the next few years this issue will become increasingly controversial in Britain.

Since the 1960s, cigarette smoking has been identified as the most serious public health problem in both the United States and Britain. The debates over smoking and health have touched upon virtually all the questions that I have suggested are generic to health promotion issues. The tobacco industry, for example, has challenged the evidence linking smoking and ill health, championed the cause of freedom to smoke and to promote its products, and portrayed smoking as a healthful, virtuous, and, during wartime, even patriotic practice. Despite persistent tobacco industry denials that a causal relationship exists between smoking and ill health, there are few who still question the conclusion reached by the U.S. surgeon general, the Royal College of Physicians, and literally tens of thousands of scientific studies, that such a relationship does exist for those who smoke.

In recent years the dominant question has been that of the nature and scope of the negative externalities imposed by smokers on nonsmokers. One can make an intellectually persuasive case that smokers impose costs on nonsmokers in the form of higher insurance rates, lost productivity, and certain public safety hazards such as fires. But these costs appear indirect, abstract, and statistical. The emergence of the passive smoking issue, and the suggestion that smokers cause physical harm to nonsmokers, has transformed the nature of the externalities claim and the policy debate. The future of policy on smoking and health, and the tobacco industry, rests, I would suggest, largely on the credibility of evidence on passive smoking. If and when that evidence is as widely accepted by policymakers and the public as is the evidence concerning the direct harm of smoking to smokers, public policy in both countries will be more coercive.

[79] See Philip Johnston, "Squeezing the Smoker," *British Medical Journal* 292 (18 January 1986): 215.

Initially both countries responded in similar fashion to the comparable public health problem posed by cigarette smoking. A situational factor, the publication of a report on smoking and health by a prestigious and credible authority moved the issue from scientific journals to the policy agenda. Within five years of the publication of these reports, both countries, despite considerable opposition from powerful economic and political interests in each, moved to alert citizens to the dangers of smoking and to inhibit the incidence of smoking.

The two countries accomplished these goals in ways consistent with their own, different, political cultures. The British, for the most part, did so through voluntary agreements, the Americans through regulation. Although I have suggested that voluntary agreements have occasioned at least as much, if not more, political conflict in Britain as regulation has in the United States, the observation is in a sense beside the point. The historically cooperative relationship between government and business in Britain, and the antagonistic relationship in the United States, would allow no other way.

British and American policy on smoking and health proceeded at much the same pace and in much the same direction until the late 1970s, when they began to diverge significantly. In the United States, anti-smoking forces, inspired by passage of the 1975 Minnesota Clean Indoor Air Act, shifted the field of battle to the states and localities and began pressing for limitations of smoking in various public settings. As measured by the number of states and localities restricting smoking, this new thrust has been remarkably successful in a relatively short period of time because the tobacco interests have not been as well represented at the state and local level as they have been in the national government. This new and more aggressive strategy was later given impetus by evidence linking health harms of nonsmokers to environmental tobacco smoke. In the United States, then, the federal structure has clearly facilitated the war on smoking by providing alternative political arenas.

In Britain no such alternatives exist. Tobacco interests have been able to maintain greater control over smoking and health policy and avoid this most recent assault on smoking. The fact that British politicians of both the right and left have been more antagonistic to limiting smoking in public places has nicely dovetailed with the structural obstacles to such a policy. Members of the Labor party have been reluctant to interfere with one of the working person's "pleasures," while "dry" Conservatives, including Thatcher, have championed the cause of allowing an informed public to be free to be foolish. British smoking and health policy continues to revolve around first-generation issues—advertising, sponsorship, and taxation. Thus far the British have relied primarily on public opinion and political persuasion to push voluntary agreements that limit the attractiveness and availability of cigarettes, and on taxation as a means of discouraging consumption. This approach will be severely tested in the area of passive smoking. It may well be that British businesses will themselves voluntarily accept the value of smoke-free workplaces and that the nonsmoking majority will insist upon smoke-free restaurants and other places of leisure. If such private approaches do not have the desired effect, and if the evidence on passive smoking becomes more convincing, there will be pressure on the government to regulate the behavior of smokers to protect the health of nonsmokers.

One critical question remains: Which nation's policies have been most successful in discouraging smoking and limiting its damaging health effects? The two most convincing and convenient measures of policy success are smoking and lung cancer rates (90 percent of lung cancer is associated with cigarette smoking). In 1960, before the link between smoking and ill health became widely known to the public, 42 percent of the adult population in the United States and 57 percent in Britain smoked cigarettes. In 1987, 29 percent of Americans and 33 percent of the British adult population smoked (see Figures 4-1 and 4-2). This represents a 31 percent decline in the United States and a 42 percent decline in Britain. Thus, although a larger percentage of the British continue to smoke, the decrease has been larger in that country than the United States.

Despite the apparent success of Britain's less coercive policy, the lung cancer rate—and one must assume the rates of other smoking-related illnesses—is greater in Britain than in the United States. In fact, the lung cancer rate in the United Kingdom is higher (about 72 per 100,000 in 1984) than in any of the other twenty-three countries of the Organisation For Economic Co-Operation and Development; the 1984 U.S. rate was 49.0 per 100,000.[80] It is not exactly clear why this is the case, although most public health experts believe that a high correspondence between smoking and alcohol drinking among British consumers is partly at fault, but the implications seem clear: more aggressive measures may be necessary to reduce the carnage in Britain associated with cigarette smoking.

The future cannot seem terribly bright for either the tobacco industry or the smoker. The industry will find it more difficult to advertise and promote its product, and smokers will find that product increasingly expensive and, for those who are not discouraged by price, difficult to use in public. The prospect for a smoke-free society in either Britain or the

[80] *Measuring Health Care, 1960–1983* (Paris: OECD, 1985), 143; and *Demographic Yearbook, 1986* (New York: United Nations, 1988). These and other lung cancer rates include trachea- and broncho- as well as lung-neoplasms.

United States by the year 2000 seems remote, but it is clear that cigarette smokers will face increased social hostility and public restrictions. As a British politician remarked not long ago, by the turn of the century smoking will be practiced only by consenting adults in the privacy of their homes.

Five

Alcohol Control Policy: Who Should Drink, When, Where, and How Much?

As an issue of public policy concern, the use or misuse of alcohol has a much longer history than that of cigarettes. In Britain, problems associated with alcohol abuse first appeared in the late fifteenth and early sixteenth centuries. Acts that sought to deal with public disorder in alehouses were adopted in 1495, 1504, and 1552. The 1552 act enforced licensing of alehouses because "the intolerable hurts and troubles to the Commonwealth of this Realm doth daily grow and increase through such abuses and disorders as are had and used in common ale-houses and other houses called 'tippling houses.' "[1] In colonial America, too, despite a generally tolerant attitude toward consumption of alcohol, there was public concern with its intemperate use. Dr. Benjamin Rush, America's foremost physician, wrote *An Inquiry into the Effects of Ardent Spirits on the Human Mind and Body* (1784), which in a later edition contained "A Moral and Physical Thermometer" detailing the vices and diseases accompanying intemperance. These included "Idleness," "Gaming," "Peevishness, Lying and Swearing," "Stealing and Swindling," "Perjury," "Burglary," "Murder," "Puking," "Dropsy," "Epilepsy," and "Madness."[2]

Given the long-standing historical concern with the public health and safety implications of alcohol abuse, one might expect a more defined set of public policies for this area than, say, for smoking. In fact, just the opposite is the case. Neither the United States nor Britain has a clearly defined policy with regard to the regulation of alcohol. Both, of course, have laws governing the sale, manufacturing, and consumption of alcoholic products. But in neither country, and most certainly not in Britain, has there been a consensus about the ultimate purpose of these discrete regulatory policies. Only recently in the United States, as the toll of alcohol-related social harm has become more visible and Americans more health-conscious, has legislative attention focused on the need for a coherent national alcohol control policy. In the United States, alcohol con-

[1] George B. Wilson, *Alcohol and the Nation* (London: Nicholson and Watson, 1940), 45.

[2] Mark E. Lender and James K. Martin, *Drinking in America: A History* (New York: The Free Press, 1982), 39.

trol policy has taken on an almost neoprohibitionist, or at least neotemperance, quality. It has manifested itself in attempts to limit the availability, accessibility, visibility, and allure of alcoholic beverages. The move toward more responsible drinking habits derives not only from the current American preoccupation with healthy life-styles, but also from a cultural predisposition, which periodically reasserts itself, toward social puritanism.

The British, for reasons developed below, have yet to adopt a common perspective, much less a coherent policy, on alcohol control. Their situation is quite different. They too have a history of temperance and prohibitionist sentiment, but it is historically more distant—reaching its apogee in the 1870s—and, measured by legislated prohibition, less successful than in the United States.[3] In fact, alcohol plays a prominent role in the social lives of many Britons. Drinking outside the home, in that quintessential and hallowed British institution the pub, is a highly preferred leisure time activity. Changing social patterns, including greater health consciousness and increased violence resulting from public drunkenness among youths, have tarnished alcohol's image somewhat, but it still enjoys a special place in British social life. In addition, Thatcher's conservatism has placed greater emphasis on unshackling the British economy, including the alcoholic beverage industry, than on restoring Victorian moral values. The result is a policy and policy environment that has actually *increased* the availability and accessibility of alcoholic beverages.

Drinking in Britain and the United States

Use, Misuse, and Abuse: Defining the Problem

Most informed citizens, scientists, and policymakers now accept the evidence that the direct use of tobacco is harmful per se; there is no such thing as a safe level of use. However, only those who advocate complete abstinence from alcohol for moral reasons would argue that alcohol per se, in any quantity, is harmful. From a public health perspective the generally accepted position, as articulated by a British member of Parliament, is that "the problem is not the product, but the abuse of it."[4] There even is some evidence that taken in "moderation" alcohol may have some beneficial effects, especially with regard to coronary heart disease.[5] For poli-

[3] See, for example, Brian Harrison, *Drink and the Victorians: The Temperance Question in England, 1815–1872* (Pittsburgh: University of Pittsburgh Press, 1971).

[4] *United Kingdom, Parliamentary Debates* (Commons) (9 May 1980), col. 766.

[5] See National Institute on Alcohol Abuse and Alcoholism, *Alcohol and Health* (Washington, D.C.: United States Department of Health and Human Services, 1981), 48–49.

cymakers, the comparison between tobacco and alcohol use is relevant for several reasons.

First, the regulation of alcohol must accommodate a wide range of drinkers, ranging from the alcoholic to the occasional social drinker. This poses the policy problem of discrimination among different types of users. The 1977 House of Commons Expenditure Committee report "Preventive Medicine" rejected higher taxation as a means of discouraging intemperate alcohol consumption because it would penalize the social or occasional drinker as well the "problem" drinker. California Congressman Vic Fazio expressed precisely the same concern in a statement opposing an increase in the federal excise tax on distilled spirits: "Those who use alcohol in excess, and do not respond to the legal and social sanctions against misuse, are least susceptible to price changes. It is the roughly 100 million consumers who drink alcohol responsibly, moderately, and legitimately who will be impacted by a tax increase."[6]

A second dilemma derives from the imprecision of the terms used to describe different levels of consumption. Discussions of alcohol consumption abound with such terms as "moderation," "light," "moderate," and "heavy" drinker, "serious drinking problem," "social drinker," "misuse," and "abuse." These labels are sometimes defined in terms of frequency of consumption, amount of consumption, consequences of consumption, or a combination of these factors. Efforts have been made to define these terms, either chemically or behaviorally, but there is little agreement within either the medical or political communities on even the most measurable dimension of the issue. A recent survey of seventy British experts on alcohol abuse sought a definition of "moderate" drinking for males. Some experts defined moderate consumption as being between 3.5 and 6.5 pints of beer a week; others felt that between twenty-eight and thirty-one pints per week was moderate.[7] The application of these rather vague characterizations is further complicated by the need to factor in such variables as age, sex, body weight, and duration of the "drinking problem." The result of this imprecision is that those who must make or implement public policy face a formidable task. According to one official in the British Health Education Council, "It's a very, very complicated educational message. . . . There's the business of what is moderation. What do you actually tell people to do? 'You should be drinking moderately' is a terribly vague educational message. What is drinking moderately?"[8]

Even in the seemingly more clear-cut case of the alcoholic, there is a great deal of diversity in terminology and definition. This problem, and

[6] *Congressional Record* (22 July 1982), 3405.

[7] *Economist*, 13 September 1986.

[8] Anthony Pollett, Health Education Council, interview by the author. Tape recording, London, England, 6 June 1984.

the limits it imposes, was acknowledged in the U.S. secretary of health and human services's fourth special report *Alcohol and Health*: "These many variations in alcohol-related problems suggest that no single concept will suffice for adequate description. Categories such as alcoholism, alcohol abuse, alcohol dependence, problem drinking, and alcohol-related disabilities are of definite although limited value. Each is useful for certain purposes and certain populations, but none can represent adequately the full range of problems associated with alcohol consumption in American society."[9] This is not to suggest, of course, that there have not been efforts to define "problem drinking," to use the broader term, or "alcoholism," to use the narrower one, but simply that they are poor guides to policymaking.

The problem of defining an acceptable level of use is further complicated by the debate over negative externalities. In testimony before the Senate subcommittee looking into alcohol advertising, the chairman of the FTC addressed this issue:

> The definition of the problem is unclear. Obviously, we do not want our children . . . consuming alcoholic beverages. Obviously, too, we want to restrain the type of third-party effects mentioned earlier—drunk driving, homicides, child abuse, and so forth. But what about consumption that affects only the one doing the consuming? Views on this differ. For what it is worth, my own view is that Government should not prevent someone from consuming alcohol, provided he or she does not hurt anyone else in the process.[10]

Yet another definitional problem confounding policymakers concerns the origins of alcohol abuse. Over the past few decades, alcoholism has been accepted as a disease rather than a measure of moral failure. Recently, evidence has emerged suggesting that at least some people may have a biological or genetic predisposition toward alcoholism. Despite widespread acceptance, the medical or disease model has not gone unchallenged. First, there are many who make an etiological distinction between voluntary and involuntary causation and place alcoholism in the former category. In this view, alcoholism is the result of personal choice. According to one British M.P., "The new generation of diseases, of which alcoholism is a good example, are caused not by nature but by our lifestyle."[11] The public policy implications of this are obvious; self-inflicted illnesses are treated differently from those over which the individual has limited, if any, control.

During the Reagan years the disease model faced a rather formidable

[9] National Institute on Alcohol Abuse and Alcoholism, *Alcohol and Health*, 29.

[10] See United States Congress, Senate, Subcommittee on Children, Family, Drugs, and Alcoholism, hearings (1985), 3.

[11] *United Kingdom, Parliamentary Debates* (Commons) (23 March 1979), col. 1969.

assault. In October 1987 the U.S. Supreme Court agreed to hear a case involving a Veterans Administration ruling that some forms of alcoholism are the result of "willful misconduct" and, therefore, are not an illness or disability. The case—actually there were two separate cases, *Traynor v. Turnage* and *McKelvey v. Turnage*—was brought by two veterans who were refused education benefits because they had not applied for them within ten years of their discharge. The two claimed that they had not taken advantage of the program because they suffered from the disabling disease of alcoholism. The Veterans Administration permits program extensions only when illnesses are not the result of "willful misconduct." It argued that "primary alcoholism," that is, alcoholism that is not the result of an underlying psychiatric disorder, was willful, irresponsible conduct.[12] In April 1988 the Court upheld the agency's position when it ruled in a four to three decision that it could continue to deny disability and other benefits to veterans disabled by alcoholism resulting from "willful misconduct." Writing for the majority, Justice Byron R. White touched upon the critical life-style issue of voluntarism: "Indeed, even among many who consider alcoholism a 'disease' to which its victims are genetically predisposed, the consumption of alcohol is not regarded as wholly involuntary." Justice White was joined by two of the Court's other conservatives, Chief Justice Rehnquist and Justice O'Connor, as well as Justice Stevens. Three of the Court's liberals, Blackmun, Marshall, and Brennan, dissented.[13]

The case is interesting for several reasons. First, it reflects the tendency of the Reagan administration, and conservatives in general, to embrace the etiological model that finds the origins of harm in individual irresponsibility. Second, it reflects a growing unease, again most prominently among conservatives, with what some see as the blurring of moral responsibility implicit in the disease model of alcoholism. According to one authority on alcoholism, "People are using alcoholism as an excuse for wife abuse, vehicular homicide [and] embezzlement." Finally, the case underscores the historical ambivalence of Americans towards alcohol and alcoholism: "As a society we still don't know whether to laugh or cry when we see someone who is drunk."[14]

Interestingly enough, although most public health professionals view the transformation of alcohol abuse from a moral to a medical issue as a positive step, it has not always facilitated the alcohol control policies that many of them would like to see adopted. Derek Rutherford, director of the Institute of Alcohol Studies in Britain, believes that the medical model

[12] *New York Times*, 25 October 1987.

[13] *Traynor v. Turnage*, 108 S.Ct. 1372 (1988), at 1383.

[14] *New York Times*, 25 October 1987.

inhibits regulation of alcohol use. He suggests that the alcohol industry, and Conservative governments, have so eagerly embraced the medical model because it has helped them to distinguish between two populations of drinkers: a small minority who suffer from a disease and require medical treatment, and the general population, who must merely be educated to drink sensibly. One need not, indeed one should not, impose regulations such as higher taxes, more restrictive licensing, and an increased drinking age on the many because of the medical problems of a few. Hence, the medical or disease model has helped mask the problems associated with irresponsible social drinking.[15]

In sum, although alcohol abuse is generally recognized as one of the most serious health problems facing both the United States and Britain, public policy must accommodate behavioral variability, definitional imprecision, and a lack of medical consensus. These are not ideal conditions for fashioning policies, particularly those that limit individual freedom and affect life-style.

"A Whole Catalogue of Harm and Damage"

The number of deaths directly attributed to alcohol abuse in each country is smaller than that caused by smoking—there are an estimated 100,000 alcohol-related deaths in the United States each year, compared to over 390,000 smoking deaths—but direct mortality from alcohol tells only part of the public health and safety problem associated with alcohol. It has been estimated that in 1990 there were about eighteen million alcoholics and people suffering from other negative effects of alcohol, and that as many as 50 to 60 million people, including family members, are the "immediate victims" of alcohol abuse. According to a 1987 Gallup poll, nearly one American family in four reported being affected by alcohol-related problems—double the proportion of 1974.[16] These problems take such forms as physical abuse of children and spouses, divorce, birth defects associated with Fetal Alcohol Syndrome, and unemployment.

One should treat estimates of the mortality and morbidity consequences of alcoholism with some caution, but it is nevertheless possible to get a sense of the enormous range of the health implications of alcohol abuse and the profound impact it has on the individual causes of mortality, what one policymaker called "a whole catalogue of harm and dam-

[15] Derek Rutherford, interview by the author. Tape recording, London, England, 31 May 1984.

[16] *New York Times*, 27 April 1987.

age."[17] Table 5-1 provides one such effort and underscores the nature of the problem in the United States.

As the table demonstrates, nearly 40 percent of all accidental and violent deaths in the United States are attributable to alcohol. Furthermore, according to Department of Justice data, alcohol is associated with almost one-half of all police arrests each year.[18] In terms of health-related problems, probably the most important indicator of chronic alcohol abuse is cirrhosis of the liver. In 1987, 26,000 people (156 per 100,000

TABLE 5-1
Estimated Number of Deaths Attributable to Alcohol, United States, 1980

Cause of Death	*Number of Deaths*	*Percentage Attributable to Alcohol*	*Estimated Number from Alcohol*
Alcohol as main cause			
Alcoholic psychoses, dependence syndrome, gastritis, hepatitis, cirrhosis, etc.	19,587	100	19,587
Alcohol as contributing cause			
Cancers of the lip, pharynx, stomach, liver, etc.	31,995	23	7,269
Other diseases			
Diabetes, influenza, hypertension, gastro-intestinal, liver	150,280	8	11,679
Accidents			
Motor vehicle, falls, fire, drowning, rail, etc.	96,987	39	37,849
Violence			
Suicide, homicide, etc.	54,499	39	21,144
Total			97,529

Source: Adapted from R. T. Ravenholt, "Addiction Mortality in the United States, 1980: Tobacco, Alcohol and Other Substances," *Population and Development Review* 99 (1984): 697-724.

[17] Sir George Young, speech before the Annual General Meeting of the National Council on Alcoholism, July 10, 1979.

[18] United States Department of Health and Human Services, *Sixth Annual Report to the United States Congress on Alcohol and Health* (Washington, D.C.: Government Printing Office, 1987), 27.

over fifteen years old) died from cirrhosis of the liver, making it the ninth leading cause of death in the United States. The direct and indirect costs to society are staggering. According to one study, the estimated economic cost of alcohol abuse and alcoholism in 1983 was $116.8 billion (see Table 5-2).

Data on the material and human costs of alcohol abuse in Britain are not as detailed as those for the United States, but those available indicate an equally dramatic impact (see Tables 5-3 and 5-4). It is estimated that in 1984, 26,500 people died from alcohol abuse (47 per 100,000 people) compared to 98,000 (41 per 100,000) in the United States. About one million people in Britain have a serious drinking problem. In terms of specific health and safety problems, just over one-half of those convicted of murder, 45 percent of those convicted of assault, and one-third of the drivers involved in traffic accidents were legally intoxicated at the time of their offences.

TABLE 5-2
Economic Costs to Society of Alcohol Abuse and Alcoholism, United States, 1983

Types of Costs	*Costs (in Millions of Dollars)*
Core costs	
Direct costs	
Treatment	13,457
Health support services	1,549
Indirect costs	
Mortality	18,151
Reduced productivity	68,582
Lost employment	5,323
Other related costs	
Direct costs	
Motor-vehicle crashes	2,697
Crime	2,631
Social welfare administration	49
Other	3,673
Indirect	
Victims of crime	194
Incarceration	2,979
Motor-vehicle crashes	590
Total	116,875

Source: United States Department of Health and Human Services, *Sixth Annual Report to the United States Congress on Alcohol and Health* (Washington, D.C.: Government Printing Office, 1987), 45.

TABLE 5-3
Alcohol Consumption and Measures of Harmful Drinking, The United Kingdom, 1970-1986

		Convictions per 10,000 Population		*Rate per 100,000 Population*	
	Consumption (Liters per Capita)[a]	*Drunkeness*[a]	*Drinking and Driving*[b]	*Cirrhosis Deaths*[c]	*Hospital Admissions for Alcohol Dependence*[d]
1970	7.03	22.3	8.40	3.71	5.17
1971	7.37	23.4	11.70	4.19	5.70
1972	7.79	24.1	13.80	4.41	6.19
1973	8.61	26.6	16.00	4.77	7.34
1974	8.85	27.7	16.30	4.63	8.11
1975	8.82	27.9	16.60	4.82	8.40
1976	9.28	28.6	14.10	4.94	8.68
1977	8.81	28.2	12.70	4.73	9.17
1978	9.50	27.5	14.10	4.97	9.13
1979	9.79	29.9	15.60	5.60	9.74
1980	9.33	30.9	18.20	5.64	10.96
1981	8.89	27.1	16.50	5.59	10.55
1982	8.86	26.1	17.00	5.42	8.91
1983	8.83	25.7	22.30	5.47	8.63
1984	9.21	22.1	22.80	5.67	9.46
1986	8.94	[e]	[e]	[e]	[e]

Source: Royal College of Psychiatrists, Alcohol: Our Favorite Drug (London: Tavistock Publications, 1986), 109.

[a] United Kingdom.

[b] Britain.

[c] England and Wales.

[d] England.

[e] Not available.

In 1985 and 1986 problems associated with alcohol misuse captured British and world attention as a result of what Prime Minister Thatcher called "appalling scenes of violence" accompanying soccer matches in Europe and Britain. At the 1985 European Cup Finals in Brussels, rioting by drunken British soccer fans resulted in thirty-eight deaths, while in another incident at Bradford City Football Grounds in England fifty-six people were killed. These events, along with other acts of violence at matches in Britain, resulted in the 1985 Sporting Events (Control of Alcohol) Bill, and a Committee of Inquiry into Crowd Safety and Control at Sports Grounds. This focused considerable public attention on the relationship between violence and the "social" use of alcohol.

Drinking in Britain has been implicated in a significant proportion of deaths or injuries due to accidents or violence: 52 percent of deaths from

TABLE 5-4
Total Social Costs of Alcohol Misuse, England and Wales, 1985

Category	*Cost in Millions of Pounds*
Cost to Industry	
Absence for sickness	723.55
Unemployment	166.74
Premature death	653.31
Household services	48.60
Cost to the National Health Service	
Hospital treatment	27.93
Alcohol-related inpatient costs	82.09
General practice costs	2.10
Cost of road traffic accidents	104.01
Cost Of Criminal Activities	18.14
Total	1826.47

Source: *Briefing Paper #1* (London: Action on Alcohol Abuse, 1987), 2.

fire, one-third of all domestic accidents and wife abuse (and divorce) cases, and 30 percent of all drownings. Deaths from cirrhosis of the liver nearly tripled over the past three decades, increasing from 1,026 in 1950 to 3,005 in 1985—from 3.4 per 100,000 population over fifteen years in 1950 to 6.9 in 1985. In 1988 it was estimated that five hundred people die in England and Wales each week from the use of alcohol. Furthermore, between 8 million and 14 million workings days are lost each year due to immoderate drinking. The litany of human misery caused by alcohol abuse is reflected in the material costs to British society. In 1985 alcohol abuse cost the British about £1.8 billion ($2.3 billion) or $41 per capita (see Table 5-4).

The human and material costs of alcohol abuse to British and American society are substantial. So too, however, are the social, cultural, and political obstacles to alcohol control policies.

Drinking and Social Values

A major difficulty confronting alcohol control policy can be illustrated by a comparison between tobacco and alcohol use. Two comments, the first by an official of the British Health Education Council, the second by the editor of *JAMA*, illustrate the difference between these areas. "We have a climate of opinion [in dealing with alcohol] that contrasts markedly with the situation in smoking. We can only dream about being in the smoking situation with alcohol. We would love to have a climate of opinion in the general population which would recognize alcohol as something out of the ordinary by way of a consumer product; not an ordinary consumer product but a psychotropic drug that has to be handled responsibly."[19] "Alcohol is a drug. It is the No. 1 drug of abuse in our society. Its only close rival is tobacco. Both are legal. One is socially acceptable in most settings; the other, less so, as this century winds down."[20]

In general there is currently wider social acceptance of alcohol than cigarette use; indeed, there is wider use of alcohol than tobacco. While cigarette smoking is now practiced only by about one-third of the adult population in each country, 83 percent of adults in Britain and 67 percent in the United States report the use of alcoholic beverages. It is not surprising, then, that it is more difficult for policymakers to impose regulations on alcohol use than on tobacco.

In addition to the cross-policy difference, there is a cross-national difference as well. As the above figures indicate, alcohol is more widely used in Britain than in the United States. Historically, the total level of alcohol consumed has been higher in the United States than Britain because of different drinking habits: more of the alcohol consumed in the United States (about 39 percent compared to 21 percent in Britain) is in the form of spirits, which has ten times the pure alcohol of beer. The difference between the two countries, however, has declined over the years (see Table 5-5). Furthermore, although consumption of alcohol has increased in both countries over the last three decades, the increase in Britain has been more dramatic; between 1951 and 1984 consumption increased 74 percent in Britain and only 34 percent in the United States. Drinking has long occupied a more prominent social and cultural role in British than in American life. As one observer put it, "Drink is taken to mark our birth, marriages and death, to seal bargains and salute the Queen. It flows through our literature from Geoffrey Chaucer to Graham Greene."[21] This descriptive analysis is supported by empirical evidence. For more than a decade a national household survey has found that the British spend more per week on "alcoholic drink consumed away from the home" than on any other leisure time activity; the closest competition came from holidays and television. Furthermore, in terms of total consumer expenditure, in 1986 there were only two things on which the British spent more money than on alcoholic drink: housing and food.[22]

It is also relevant to note that although the British drink more than

[19] Anthony Pollet, interview by the author. Tape recording, London, England, 6 June 1984.

[20] George Lundberg, "Ethyl Alcohol—Ancient Plague and Modern Poison," *JAMA* 252 (12 October 1984): 1911.

[21] *Times* (London), 18 August 1981.

[22] See United Kingdom Central Statistical Office, *Social Trends 1986* (London: Her Majesty's Stationery Office, 1986): 169.

TABLE 5-5
Per Capita Consumption of Alcohol, United States and Great Britain, 1951-1986 (In Liters)

	United States	*Great Britain*
1951	7.59	5.29
1955	7.56	5.40
1960	7.82	5.77
1965	8.58	6.52
1970	9.56	7.11
1975	10.16	9.39
1980	10.47	10.76
1984	10.17	9.20
1986	9.99	not available

Sources: Institute of Alcohol Studies, *Alcohol Statistics* (London: ND), 2; *Economist*, 13 September 1986, 65; United States Department of Health and Human Services, Fourth and Sixth *Annual Report to the United States Congress on Alcohol and Health* (Washington, D.C.: Government Printing Office, 1981, 1987).

Americans, they do not view drinking as much of a personal or national problem. For example, Gallup found that only 10 percent of the British, compared to 21 percent of Americans, reported that alcohol was a cause of trouble in their families. In addition, in response to a question on the seriousness of alcoholism in their country, 71 percent in the United States responded that it was a "very serious" problem, whereas only 44 percent in Britain did so.[23] These different perceptions, despite the comparability of the actual problem, is probably the result of "a prevailing over-permissive attitude to alcohol" in British society.[24]

The policy consequences of the sociocultural role of alcohol in British society have been profound: "One of the main reasons why it is almost impossible to hold a rational debate about alcohol is because it is embedded in our culture from the humblest to the most exalted level."[25] Not only rational debate but policy options as well are circumscribed by prevailing social norms. In its most general form the impact is seen in the overall approach to alcohol control policy. As a study by the Department

[23] *Gallup Report No. 242*, November 1985, 68–69.

[24] Derek Rutherford, "The Alcohol Problem in Britain," paper presented at the Seminar on the Medico-Social Risks of Alcohol Consumption, Luxembourg, November 16–18, 1977.

[25] *Times* (London), 18 August 1981.

of Health and Social Security put it, "In a society where strong social controls against alcohol misuse exist and are generally accepted, measures of the first kind [i.e., health education] might prove sufficient; however in societies like our own, where social controls are less strong, health education alone has not been effective."[26]

American cultural attitudes concerning alcohol generally are not nearly as well defined as British attitudes. One recent study, *Drinking in America*, characterized our time as "an age of ambivalence." "American ambivalence toward the subject [of alcohol] is undeniable. Perhaps 30 per cent of the nation's citizens do not drink at all, yet others tolerate considerable latitude in drinking behavior. And although the nation spends huge sums annually in consumer purchases of beverage alcohol, and generally accepts the idea that alcoholism is a disease, Americans still place a severe social stigma on alcoholism."[27] Yet there are others who see America as having entered an era of "neotemperance." One of the authors of *Drinking in America*, James Kirby Martin, who in 1982 had described America's "ambivalence," in 1985 saw a new cycle of opinion emerging. "It's pretty clear to me that we have entered a kind of neo-temperance, neo-Prohibitionist phase of activity in the United States, especially in the last five years. I think we're in a long-term anti-alcohol cycle."[28] The rekindling of anti-alcohol sentiment in the United States is anchored, as it has been since the nineteenth century, in Bible Belt, Protestant fundamentalism. It may be that this most recent temperance cycle owes its vitality, in part, to the revival of fundamentalism in the United States.

Evidence of this new phase can be seen in both self-reported drinking behavior and actual consumption. In national surveys in 1971 and again in 1985 Lou Harris asked: "Compared with five years ago, are you drinking more, less, or about the same?" In 1971, 31 percent of the respondents indicated that they were drinking less; in 1985, 45 percent reported a decline in drinking.[29] There is empirical evidence that alcohol consumption has declined: in 1986 U.S. annual per capita consumption of alcoholic beverages was 2.58 gallons of pure alcohol, down from the 1982 peak of 2.77 gallons per person. In 1988 the hard-liquor industry reported its lowest level of beverage shipments since it began its survey in 1970. Whether it is because of a historical ambivalence or resurgent neo-

[26] United Kingdom Advisory Committee on Alcoholism, *Report on Prevention* (London: Her Majesty's Stationery Office, 1977), 4.

[27] Lender and Martin, *Drinking in America*, 191.

[28] *New York Times*, 1 October 1985. For similar views on a neo-temperance attitude in the United States, see *Gallup Report No. 242*, November 1985, 5; and *Business Week*, 25 February 1985, 112–13.

[29] *Business Week*, 25 February 1985, 113.

temperance, behaviorally and attitudinally Americans differ from their British cousins with regard to the role of alcohol in society.

Alcohol Abuse and Public Policy

There are two policy approaches to the problems associated with alcohol abuse, which stem from the two major etiological explanations of the problem. The first is a medical approach, the second, an alcohol control approach. The first focuses on the alcoholic and the severe problem drinker. It accepts alcoholism as a disease, involving a chemical addiction, perhaps even a genetic predisposition, and addresses it in a traditional medical manner. Provisions are made to help alcoholics obtain medical attention, and funds are allocated for research on the causes and cures of the disease. Although it is recognized that alcoholics contribute to the public safety problems discussed above, such as traffic fatalities, family violence, and homicides, their behavior is viewed differently by policy-makers and others than that of those who cause society harm through what are perceived to be voluntary lapses in social judgment—drinking too much before driving, or abusing a child or spouse while drunk. This difference was acknowledged recently by the Royal College of Psychiatrists: "The fact remains that at the extreme end of the spectrum there are people who are suffering from a type of involvement with alcohol which requires very special understanding. These are people who have developed the dependence syndrome—who are addicted to, or dependent on, alcohol."[30] The social consequences of the behavior of an alcoholic or an irresponsible social drinker may be the same, but from the perspective of public policy-making one is seen as a disease requiring a medical response, and the other is viewed as a socially harmful act of self-indulgence requiring a political response. It is the latter problem and perspective upon which I will be focusing.

Because it is assumed that alcohol itself, if used wisely, is a safe product, public policy is geared toward preventing overindulgence of alcohol, not eliminating its use. Broadly speaking, alcohol control policies aim to influence the level and manner of consumption. This can be accomplished in one of three ways: by setting constraints on consumption through regulating price, condition of sales, and drinking age; by penalizing demonstrably dangerous behavior such as drinking and driving; and by creating favorable social controls through education and information about alcohol misuse (for example, through advertising).

[30] Royal College of Psychiatrists, *Alcohol and Alcoholism* (London: Tavistock Publications, 1979), 37.

Limiting Consumption, Limiting Harm

SETTING PRICE: TAXATION

It is generally accepted within the medical and health promotion communities that there is a strong relationship between the overall level of alcohol consumption in a society and the amount of harmful drinking. This hypothesis, which has its theoretical origins in the work of a French statistician, Solly Ledermann (hence "the Ledermann hypothesis"), posits that incremental changes in consumption move people in and out of a drinking "danger zone." The higher the level of aggregate consumption in a society, the higher the level of harmful drinking.[31] The posited relationship between consumption and alcohol harm is vital to policy-making in this area. One British M.P. summarized this association and adumbrated policy strategy in this way: "It shows that alcoholism is not caused so much by problems at work, unhappy marriages, by the stress of modern life or by mothers-in-law. It is caused, quite simply, by alcohol."[32] In this view, the magnitude of the public health problem posed by alcohol is directly related to levels of consumption.

Although some proportion of problem drinking may emanate from predisposing biological factors, the prevailing wisdom in both countries is that most of the problem is self-inflicted and therefore essentially a political, not medical, problem. Given the documented relationship between consumption levels and harm, a major goal of health advocates has been to reduce alcohol use. And the two factors that most influence consumption are price and availability.

Of these two factors, price is generally held to be the more important one, and in terms of public policy taxation is the key to affecting price. To begin with, it should be noted that alcohol consumption is more price-sensitive than tobacco consumption, at least for the nonalcoholic. One study has shown that even a one-percent real price increase would lead to reduced consumption.[33] It is therefore important to note that cigarette tax policy, especially in Britain, has reflected increased sensitivity to the health issue, but alcohol policy has not. In neither the United States nor Britain have taxes been enlisted in the war on harmful drinking habits; alcohol taxes serve exclusively as revenue- not health-enhancing policy tools. Furthermore, although taxes on alcohol have risen slowly over the years in both countries, these increases have not kept pace with inflation. As a result, the real price of alcoholic beverages has declined over time,

[31] Ibid., 93.

[32] *United Kingdom, Parliamentary Debates* (Commons) (23 March 1979), Col. 1971.

[33] Tony McGuiness, *An Econometric Analysis of Total Demand for Alcoholic Beverages in the U.K., 1965–75* (Edinburgh: Scottish Health Education Unit, 1979), 13.

and consumption has increased. According to a report prepared by the Central Policy Review Staff, a British government "think tank," "a major factor in increased total consumption since the war has been the relative cheapening of alcohol" due to tax policies.[34]

In the United States, the slower rise in federal alcohol excise tax is reflected in the diminishing proportion of national revenues realized from alcohol taxes. For example, the federal excise tax on beer and wine has remained the same since 1951, and the excise tax on a fifth of whiskey has increased only forty cents over the same period. As a result, the proportion of federal budget receipts from alcohol taxes has declined dramatically. In 1941 alcohol taxes provided about 11 percent of federal revenues; in 1988 they accounted for less than one percent. When expressed in constant dollars, federal excise taxes on beer, wine, and liquor are all lower now than they were fifty years ago, at the time of the repeal of Prohibition.[35]

The situation is much the same in Britain. Taxes on wine, beer, and spirits have increased over time, but in no case have these increases equaled or exceeded the overall Retail Price Index increase. Hence, the relative cost of alcoholic beverages has declined. In 1950 a male worker had to work twenty-three minutes to earn the price of a pint of beer, and six and a half hours to pay for a bottle of whiskey; in 1984 it took just twelve minutes work to buy a pint of beer, and two hours and sixteen minutes to buy the whiskey.

It is with the relationship between price and consumption in mind that various health advocates in both countries, including the Royal College of Psychiatrists, the BMA, and, in the United States, the Center for Science in the Public Interest and Surgeon General Koop, have urged that taxes on alcohol be increased at least to keep pace with inflation. Yet it is precisely because of the consumption-price relationship that taxes have not increased more rapidly. Derek Rutherford of the British Institute for Alcohol Studies argues that "the government is more dependent on alcohol than any alcoholic."[36] In recent years about 15 percent of British expenditure taxes have come from alcohol taxes and duties. This proportion is considerably lower than the nearly 40 percent of all revenues

[34] United Kingdom Central Policy Review Staff, *Alcohol Policies* (London: 1979), 19. The author's citation is to the confidential, unpublished version of this report that the British government refused to publish. The report was published subsequently in Sweden as Central Policy Review Staff, *Alcohol Policies in the United Kingdom* (Stockholm: Sociologiska Institutionen, 1982).

[35] National Alcohol Tax Coalition, *Impact of Alcohol Excise Tax Increases on Federal Revenues, Alcohol Consumption, and Alcohol Problems* (Washington, D.C.: National Alcohol Tax Coalition, 1985), 2.

[36] Alex Paton, "The Politics of Alcohol," *British Medical Journal* no. 290 (January 5, 1985): 1.

supplied by alcohol in 1900, but it still represents a substantial amount (about $7.6 billion in 1985). Furthermore, about two percent of all British export earnings come from alcohol. Finally, it is important to remember that alcohol, like tobacco, is part of the RPI; hence price increases here too have potentially broad implications for inflation and benefits programs.

The alcoholic beverage industry represents a $70 billion a year enterprise in the United States. The industry produces about $13 billion annually in state and local taxes, and even though they comprise a much smaller proportion of public revenues than in the past, they are still large enough to have considerable influence on public policy. Governments, in short, have been reluctant to kill this other goose that lays the golden egg.

The economic consequences of decreased alcohol consumption are not limited to lost government revenues. The United Kingdom is the world's second largest exporter of alcoholic beverages, accounting, in 1980, for nearly 22 percent of total world exports—France was first with 30 percent. In that year, four of the top ten alcoholic beverage corporations—Imperial Group, Grand Metropolitan, Lonrho, and Allied-Lyons—were headquartered in the United Kingdom.[37] It is not surprising, then, that the alcohol industry is a major employer in Britain: an estimated 750,000 people are directly employed in the production, distribution and sale of alcohol products. They make up about 1.5 percent of the total workforce compared to 2.0 percent in the United States. This figure does not include those in the advertising industry or the media who earn considerable sums from alcohol promotion. As a result, the British alcohol lobby, including more than sixty M.P.s with direct affiliation to the industry, representing distillers, 117 breweries, 80,000 British pubs, and a workforce of three-quarters of a million people, is one of the strongest and most influential in the country. Its influence is enhanced by virtue of its historical alliance with agricultural interests, for example, hops and barley growers.

The dilemma facing Conservative and Labor governments alike was captured by the Central Policy Review Staff: "A Government's desire for revenue, its responsibility for economic policy generally and for the prosperity of the [alcohol] industry, and its sensitivity to the political factors involved are realities against which recommendations for [policy] change must be set."[38] In 1981 the Department of Health and Social Security published a discussion document, "Prevention and Health: Drinking Sensibly." In the document, which was highly criticized by health promotion advocates, the department acknowledged these realities and failed to en-

[37] John Cavanagh and Frederick F. Clairmonte, *Alcoholic Beverages: Dimensions of Corporate Power* (New York: St. Martin's, 1985), 22, 45.

[38] United Kingdom Central Policy Review Staff, *Alcohol Policies*, 14.

dorse recommendations by the Royal College of Psychiatrists, the Review Staff report, and the BMA to halt the decline in the real price of alcohol and use tax policy to moderate consumption.

CONTROLLING AVAILABILITY AND ACCESSIBILITY

Alcohol consumption is affected not only by price but by the availability and accessibility of the product. In this area cultural differences between the two countries are most obvious. In the United States a number of trends have conspired to reawaken the dormant neotemperance sentiment that has long been a feature of the American moral and cultural legacy. These trends include the seemingly singular preoccupation with healthy life-styles among a large segment of the population, the revitalization of both religious fundamentalism and political conservatism that traditionally has included a heavy dose of moral puritanism, and changing health problems among America's young people, especially those associated with drug and alcohol abuse. As a result, U.S. policy has moved to limit both the availability and accessibility of alcoholic beverages. The British, on the other hand, despite similar public health patterns, and indeed similar conservative political but not religious trends, have *increased* availability and accessibility in recent years. In this issue area, social attitudes toward drinking, not political ideology or structure, have determined public policy.

The United States: Selective Prohibition

Recently a number of U.S. states have acted to discourage overall consumption of alcoholic beverages in drinking establishments. Led by New Jersey and Ohio, several states have banned so-called "happy hours" and other promotions designed to attract patrons and increase consumption. In 1983 the Presidential Commission on Drunk Driving recommended that states adopt "dram shop" laws that establish liability against any person who sells or serves alcohol to an individual who is visibly intoxicated. By 1985, twenty-two states had adopted such laws. But the most dramatic and revealing effort in recent years to limit access to liquor was the 1984 federal law that denied federal highway funds to states with minimum purchasing or drinking ages of less than 21 years. In effect, this law has created a national minimum drinking age. The policy debate over the drinking age reveals much about the politics of health promotion in United States.

Just over 47,000 people were killed in motor-vehicle accidents in 1988; 23,352 of these deaths were alcohol-related. Each year a disproportionate number of these alcohol-related deaths involve young people, especially young males. In 1987 drinking and driving accidents were the number-one killer of teenagers: the alcohol-related highway fatality rate was 19 per 100,000 people for those under twenty-one and 11 per 100,000 for those over twenty-one. People under twenty-one years old accounted for only 8 percent of the driving population but 17 percent of the alcohol-related fatalities.

In April 1982, in response to what he later called "a national menace, a national tragedy, a national disgrace," President Reagan appointed a Presidential Commission on Drunk Driving. The commission issued its report in November 1983, and among its recommendations was the establishment of a nationwide minimum alcohol purchasing age of twenty-one on a state-by-state basis. The commission pointed to existing evidence that demonstrated that raising the legal drinking age resulted in declining fatalities, especially among young males.[39] It recommended denying federal highway construction funds to those states (numbering twenty-seven in 1984) that did not subsequently adopt a twenty-one-year old minimum.

Federal encouragement of stricter state efforts against drunk driving began in earnest in 1981 when, at the prodding of Mothers against Drunk Drivers (MADD), Congressman Michael Barnes (D-Maryland) and Senator Clairborne Pell (D-Rhode Island) introduced legislation that urged states to adopt mandatory minimum penalties for drunk driving offenses, including automatic and mandatory suspension of licenses, participation in alcohol education programs, and jail terms. In 1981 Congressman Barnes and others called upon President Reagan to establish a presidential commission on drunk driving. Responding to congressional and public pressure, the president appointed the commission in April 1982.

With an aroused public to spur it on, Congress responded with a variety of legislative proposals, including incentive grants to encourage stricter state laws against drunk driving and the creation of a National Driver Register that would enable states to exchange information on the driving records of individuals. This latter proposal ultimately became law in July 1984. In addition, in 1982 and 1983 Congressman William Goodling (R-Pennsylvania) introduced concurrent resolutions urging adoption of a minimum drinking age of twenty-one. In April 1983 Congressman John Porter (R-Illinois) introduced a bill to prohibit the use of federal highway funds in any state in which the minimum drinking age was under twenty-one. Although the House originally adopted the measure as part of highway-transit legislation, it was deleted in the Senate. There was, then, considerable public awareness of, and state and federal

[39] United States Presidential Commission on Drunk Driving, *The New Hope of Solution* (Washington, D.C.: 1983), 10.

legislative sensitivity to, the issue of drunk driving among young people when the presidential commission issued its report in November 1983.

With an important boost from the commission report, MADD, with the support of the National Council on Alcoholism, the National Safety Council, the National Parent-Teachers Association, and the AMA, undertook an enormously compelling and successful media and grassroots lobbying campaign to win legislative approval of a bill that would establish a national drinking age of twenty-one. On June 7, 1984, a purchasing-age amendment was added in the House to an interstate highway funding bill. The amendment, which was approved by a vote of 297 to 73, provided for withholding federal highway funds from those states that failed to increase the purchasing age to twenty-one years. Senate action on the bill was delayed over provisions unrelated to the drinking-age issue, but eventually the provision was attached to a noncontroversial House bill requiring the states to use some federal funds to promote the use of automobile child restraints. On June 26, 1984, the Senate, by a vote of 81 to 16, approved a bill providing that beginning in October 1986, the government would withhold a proportion of highway funds—5 percent in 1987, 10 percent in 1988—from noncomplying states.

Despite the considerable popularity of the issue—a Gallup poll at the time found 79 percent of all age groups and 61 percent of those eighteen to twenty-four years old in favor—and the brilliant lobbying effort by MADD, several members of Congress and President Reagan voiced opposition to the measure. Reagan initially rejected his own commission's recommendation because he viewed it as another example of coercive federal meddling in state affairs. The president abandoned his opposition following the House vote largely for political reasons; 1984 was a presidential election year, and the drinking-age issue was too popular to ignore. Others in Congress, however, shared the president's states' rights and Big Brother concerns. Senator Max Baucus (D-Montana) took a position shared by many of his western-state colleagues: "Montanans are an independent people. When they have problems, they like to solve them on their own. They like to make their own choices. They don't like to be pushed and they don't like to be maneuvered. Most of all, they don't like to be blackmailed by their 'Big Brother' in Washington."[40] Other members voiced concern about the "slippery slope" upon which Congress was entering. Congressman James Jeffords (R-Vermont) warned after the Senate vote, "Last night they [eighteen-to-twenty-one-year-olds] lost the privilege to drink as far as Congress is concerned. Who knows what they will lose on the eve of the next recess."[41]

[40] *Congressional Record* (26 June 1984), 8213.

[41] Ibid. (28 June 1984), 7433.

Despite this opposition, the raising of the drinking age had simply become too popular an issue to stop. The speed with which the bill passed, and the magnitude of the support in both houses, took even its supporters by surprise. One Republican congressional aide explained the strong support in this way: "Politically, some issues become apple pie. This was an apple pie issue."[42] He might have added that when the apple pie is baked by a group of parents who have lost their children in drunk-driving accidents, and who showed remarkable skill in selling that pie, it becomes an irresistible product. Senator Alan Simpson (R-Wyoming), an opponent of the bill, paid homage to MADD during the Senate debate when he referred to "this significantly impressive group . . . and they are indeed, or we would not be here on the floor in this fashion and this swiftly if they were not effective."[43] The bill was the first major piece of federal legislation dealing with drinking since the end of Prohibition in 1933. President Reagan, with Candy Lightner, founder and head of MADD, in attendance, signed the law on July 17, 1984.

The battle over the drinking age was, of course, not over. There were still twenty-seven states that allowed drinking and purchasing of alcohol by those under twenty-one years old. Over the next two years state legislatures debated this issue with concerns over states' rights and federal Big Brotherism looming especially large. Legislators from states as diverse as New York and Idaho charged the federal government with "blackmail" and holding the states hostage.[44] But as was the case at the national level, the work of the supporters of the raised drinking age, along with strong public approval and the threat of the loss of federal highway funds, moved states to change their laws. On October 1, 1986, the deadline for loss of federal funds, only eight states were still not in compliance, five of them—Colorado, Idaho, Montana, South Dakota, and Wyoming—in the West, where political independence, rugged individualism, and disdain for Big Brother in Washington are enshrined in the local political culture.

Opponents of the change, including the alcoholic beverage producers, restaurant, hotel, and tavern owners, and some college student groups, saw one last chance to stop the march toward what they believed to be selective prohibition. Following passage of the law, the attorney general of South Dakota filed a suit against Transportation Secretary Elizabeth Dole, challenging the constitutionality of the federal law. The suit claimed that the Twenty-first Amendment, which repealed Prohibition, left it up to the states to set alcohol control policies. In December 1986, following rejection of the claim by federal district and appeals courts, the U.S. Su-

[42] *New York Times*, 1 July 1984.

[43] *Congressional Record* (26 June 1984), 8224.

[44] New York Assembly, 17 June 1985, 24; Portland *Oregonian*, 10 February 1987.

preme Court agreed to hear the case, *South Dakota v. Dole*. South Dakota's appeal was supported by the eight other states as well as the National Beer Wholesalers Association. On June 23, 1987, in a seven-to-two decision, the Court upheld the 1984 law. In the majority opinion, Chief Justice Rehnquist said that the law's purpose of reducing drunk driving was "directly related to one of the main purposes for which highway funds are expended: safe interstate travel."[45] By the time the case reached the Court only six states still had not complied. One of these, South Dakota, earlier had passed a law raising the drinking age to twenty-one, but with the provision that it would take effect only if the state lost its case in the Supreme Court.

In March 1988 the last holdout, Wyoming, whose legislature had defeated four bills to raise the drinking age, finally, and narrowly, became the fiftieth state to adopt a drinking age of twenty-one. The state could no longer afford the loss of $8 to $10 million a year in federal highway funds. In addition, since it had the only drinking age of eighteen in the country, it was attracting young people from adjacent states who were drinking and driving home drunk with predictable results; the state was becoming known as "bloody Wyoming."

Great Britain: Liberalizing Hours

The attitude toward controlling alcohol problems by regulating the accessibility of alcoholic beverages is different in Britain. To begin with, the legal drinking age is eighteen years, and there is virtually no chance that it will be raised. Even militant health promotion interests do no advocate increasing the drinking age. A 1972 Report of the Departmental Committee on Liquor Licensing, the Erroll Report, even recommended lowering the drinking age to seventeen. The recommendation was rejected by the Labor government, but support for it continues among some members of Parliament and alcohol-manufacturing and sales interests. Most public officials admit that the law is largely ignored and that the public houses routinely serve 16- and 17-year-olds. In 1986 a government study, *Adolescent Drinking*, disclosed that 40 percent of the 16-year-olds surveyed admitted to drinking illegally in pubs.[46]

Second, the British have actually liberalized access to alcohol. The first effort along these lines was the 1961 Licensing Act that, among other things, allowed alcoholic beverages to be sold in supermarkets, retail stores, and gasoline stations. These outlets proliferated from 29,700 in 1973 to 43,891 in 1983. But the most revealing facet of the debate over accessibility involves the question of hours of sale in pubs. Licensing laws, which until 1988 limited the number of hours and times of the day that

[45] *New York Times*, 24 June 1987.

[46] *Times* (London), 18 December 1986.

pubs could be open (11:00 A.M. to 2:30 P.M., and 5:00 P.M. to 11:00 P.M.), originated during World War I, when Lloyd George proclaimed, "We are fighting Germany, Austria and the drink." Under the Defence of the Realm Act (1914) drinking hours were regulated around important harbor areas and within the armed forces. In the following year the Central Control Board was established to regulate the hours of sale to workers in strategic industries such as munitions, although ultimately the regulations were extended to the general population.

Licensing hours have long been a matter of corporate complaint in Britain. The alcohol-manufacturing and sales industries, as well as tourist interests, argued that the law hurt their businesses. Although the Erroll Committee recommended extending hours, the government did not act upon the recommendation. Renewed interest in and impetus for changing the licensing laws came in 1976 as a result of a Scottish Licensing Act. The act grew out of a recommendation by the Clayson Committee, the Scottish version of the Erroll Committee, that pub hours be extended beyond eight and a half hours per day. The Clayson Committee recommended hours of 11:00 A.M. to 11:00 P.M. The government did not accept this recommendation, but it did extend closing time by one hour and allow pub owners to apply to licensing boards for further extensions. Such applications were frequently requested and readily granted. As a result, more than 60 percent of the pubs in Scotland are open beyond the standard hours, most commonly from 11:00 A.M. to 11:00 or 12:00 P.M.

By the early 1980s, supporters of liberalizing "permitted hours" could now point to the Scottish experience which showed that contrary to the apocalyptic predictions of many health advocates there was no dramatic increase in alcohol-related problems from liberalization. Actually, government data comparing the periods of 1979–1983 to 1971–1976 showed a 13.6 percent decline in drunkenness offenses. Over the two same time periods, drunk-driving offenses in Scotland increased by only 1.2 percent compared to 36 percent in England and Wales.[47] Furthermore, the *British Medical Journal* reported that there had been no change in the alcohol abuse problem in Scotland since liberalization. Opponents of extended hours challenged these conclusions by pointing out that the studies did not take into account either the impact of the recession of the early 1980s or the 1980 Criminal Justice (Scotland) Act that made it unlawful to consume alcohol on the way to sporting events. This law has been credited with a decline in arrests for public drunkenness since 1980.[48]

The campaign to revise the licensing laws began in earnest in 1984 with

[47] *Times* (London), 26 May 1985.

[48] *Economist*, 4 July 1987.

the creation of a lobbying group called "Flexi-Law Action Group," or "Flag," comprised of fourteen organizations representing brewers, distillers, hotel, restaurant, tourist, and public-house operators, and interest groups. "Flag" proposed allowing pubs to open for any twelve-hour period between 10:00 A.M. and midnight. Over the next two years several studies appeared in support of extended operating hours. These studies argued that liberalization would result in thousands of new jobs—estimates ranged from 15,000 to 65,000—and would reduce alcohol-related problems because there would no longer be the problem of last minute "beat-the-clock drinking" when time was called in pubs.[49] It was, however, the economic implications of expanded pub hours that attracted support from the Thatcher government. In addition to the prospect of increased jobs, more investment in the leisure sector, and increased revenue from tourism, there were the more basic enticements of further economic deregulation and increasing "consumer choice."

Although little time remained for the government to pursue liberalization in the waning days of the second Thatcher term of office the Government announced in December 1986 that it would support a proposed private member's bill allowing "all day" opening for pubs. The bill came up for a second reading on January 30, 1987, but opponents were able take up enough time to prevent a vote. The Conservatives, however, included a commitment to liberalize pub hours in their manifesto for the June 1987 general election. On June 25, 1987, following her party's reelection, Thatcher included in her program, read by the Queen, a plan to introduce a bill permitting pubs in England and Wales to stay open up to twelve hours a day. Typically British prime ministers get what they want, but opposition to extended hours, even on the Tory benches, promised a lively parliamentary debate. Certainly the policy ran contrary to the direction of alcohol control policies in the United States and elsewhere. This point was made in the House of Commons by Sir Bernard Braine, one of the most respected spokesmen on alcohol policy in Parliament. "Why," Sir Bernard wondered, "when other countries are desperately tightening up on the availability of alcohol . . . are we taking up parliamentary time relaxing it?"[50] Despite opposition from health promotion interests, in April 1988 the House of Commons overwhelmingly approved a Licensing Bill extending pub hours from 11:00 AM to 11:00 PM Monday through Saturday. The bill went into effect in August 1988.

The contrast between the two countries could not be greater. In the United States neotemperance sentiment prevailed in the face of a reluctant

[49] See John Lewis, *Freedom to Drink* (London: Institute of Economic Affairs, 1985); *Time to Call Time* (London: Adam Smith Institute, 1986); Michael Colvin, *Time Gentlemen Please* (London: Conservative Political Centre, 1986).

[50] Sir Bernard Braine quoted in *British Medical Journal* no. 295 (11 July 1987): 154.

conservative president and considerable states' rights opposition to impose "selective prohibition." In Britain, where much the same evidence linking availability, consumption, and harmful drinking habits was accepted by the health promotion community, the cultural obstacles not only prevented greater alcohol control, but actually led towards liberalizing access.

Controlling Irresponsible Behavior

DRINKING AND DRIVING IN BRITAIN

This is not to suggest that the British have been unmindful of the problems associated with alcohol abuse in general—in October 1987 an interdepartmental group was created to examine responses to alcohol misuse—or the problem of drinking and driving in particular. In addition, although the recent attention given to alcohol-related traffic fatalities and injuries has focused on young people, the problem extends beyond this age group. Over the last twenty-five years more than one-third of all drivers killed in traffic accidents have had alcohol levels above the prescribed legal limit. Certainly no policymaker in Britain champions the cause of drinking and driving. Yet given the prominent cultural role played by "going down to the pub for a pint," dealing with this public health problem has proven to be particularly thorny.

The policy debates in Britain over drinking and driving have often revolved around two technical issues: defining drunkenness and determining what are technically and politically acceptable measures of it. The two issues are obviously related, but they have not always been treated in tandem. They were joined, however, in the 1967 Road Safety Act. In that law the Labor government did two things no previous government had been willing to do: it introduced a specific legal blood-alcohol limit for drivers, and it required the use of breath analysis to determine impairment.

Until 1967 Great Britain did not have any specific legal blood-alcohol limit for drivers. Indeed, the Conservative government rejected an amendment to the 1962 Road Safety Bill that would have established such a formal limit. The argument offered then, and in previous years, was that impairment from alcohol varied from individual to individual depending upon physical build, weight, and overall health. In addition, establishing a specific limit, which was often illustrated by a number of whiskies or pints of beer, would encourage people to drink up to that limit, which in some cases might be more than was prudent.

Another problem raised in the 1962 debate was the difficulty in measuring blood alcohol. There were three kinds of tests: blood, urine, and breath. The under-secretary of state for the Home Department noted that it was "repugnant" to many to make compulsory either blood or urine analysis, and that at the time there was no breath-testing instrument reliable enough to be accepted as evidence in court. In any event, the proposal for a specific limit was rejected.

Thus, until 1967, conviction for being either "unfit to drive" or "for the time being impaired" was based on the clinical judgment of a police physician or testimony of witnesses, but not necessarily chemical analysis, because blood, urine, and breath tests were voluntary. Police complained that convictions were difficult to obtain because of the highly subjective nature of the clinical evidence and the reluctance of juries to return guilty verdicts. With driving home from the local pub so prevalent in Britain, there seemed to be widespread feeling on the part of many jurors that "there but for the grace of God go I." As a result, the police tended to press action only in the most egregious cases. By 1967, however, the accumulated evidence on the contribution of alcohol to road deaths and injuries (estimated in 1966, a particularly bad year, to be 40 percent), along with the development of more reliable breath analysis tests and a change in government, prompted more vigorous measures to discourage drunk driving.

The first step was to make it illegal to drive with a blood alcohol concentration (BAC) of 80 milligrams per 100 milliliters or over, or a urine alcohol concentration over 107 milligrams per 100 milliliters. (By comparison, most U.S. states have a BAC limit of 100 milligrams; only Oregon and Utah have 80 milligram limits. Shortly before he left office in 1989, Surgeon General Koop recommended lowering the BAC to 80 milligrams in all states immediately, and to 40 milligrams by the year 2000.) Second, the law authorized the police to carry out breath analyses of those suspected of driving under the influence of alcohol, even though the breath test itself could not be used as evidence of impairment. The breath test was in effect a screening process. Refusal to take a breath, urine, or blood test constituted a punishable offense as well as evidence for the prosecution. Adoption of the "breathalyser" test, which was introduced by Labor Transport Minister Barbara Castle—and dubbed "Barbara's Breathalyser"—"caused great resentment amongst Labour's grass roots." The measure, along with an unsuccessful proposal to ban cigarette coupons, led Prime Minister Harold Wilson's leader of the House of Commons, Richard Crossman, to lament that, "We're in danger of becoming known as the government which stops what the working classes really want."[51]

[51] See Peter Taylor, *Smoke Ring: The Politics of Tobacco* (London: The Bodley Head, 1984), 84.

The initial impact of the drinking-related provisions of the 1967 Road Safety Act was dramatic. Drunk-driving convictions nearly doubled between 1966 and 1968, from 9,590 to 18,374, while traffic fatalities involving legal intoxication dropped from about 30 percent of all fatalities to 17 percent. Unfortunately the decline, while highly celebrated, was also short-lived; by the mid-1970s alcohol-related traffic fatalities exceeded pre-1967 levels. Part of the blame for the increase was attributed to lackadaisical law enforcement, and part to technical problems with the law. In any event, in 1974 the newly elected Labor government appointed a special commission on drinking and driving named after its chairman, Frank Blennerhassett.[52]

The Blennerhassett Commission issued its report in 1976 and made six major recommendations. The most important and controversial dealt with improving the deterrent effect of the drinking law. Specifically, the commission proposed discretionary or random testing of motorists. It felt that the prospect that any drivers might be stopped, not simply those who displayed the most obviously erratic behavior, would be the strongest deterrent to drinking and driving. Second, and again with deterrence in mind, it urged that high-risk (that is, repeat) offenders be subject to special procedures such as demonstration to a court that their drinking habits had changed and that they no longer presented a danger to society before their licenses would be reinstated. Third, the commission recommended that administration of the law be simplified to reduce the number of cases dismissed because of procedural errors. Fourth, it urged retention of the existing BAC level of 80 milligrams per 100 milliliters, despite the fact that only one other Western European country, the Republic of Ireland, had a higher limit, and most had lower limits. Finally, the commission recommended that breath-testing in place of blood or urine analysis be used to determine BAC levels.

The Labor government generally accepted the recommendations, although it had reservations about random testing, but it did not do so with sufficient enthusiasm to submit changes to Parliament. As a confidential report noted, "It was not thought appropriate or possible to find parliamentary time for the legislation."[53] The Labor government had not taken any action on the Blennerhassett recommendations by the time it left office in May 1979.

The Thatcher government thus inherited the problem of what to do about a serious public health problem. Since the mid-1970s the number

[52] United Kingdom Department of the Environment, *Drinking and Driving* (London: Her Majesty's Stationery Office, 1976).

[53] United Kingdom Central Policy Review Staff, "Alcohol Policies," *Confidential Report* 1 (1979): 45.

of drivers killed in traffic accidents with BAC levels above the legal limit continued to hover around one-third of all deaths despite increased health education efforts. In 1982 the Thatcher government adopted one of the key Blennerhassett recommendations, the use of breath analysis as the primary measure of BAC. On the other hand, the government rejected other key provisions. It said it would study the proposal for strict disqualification of repeat offenders, but as of 1989 it had not taken any action on it.

It is the random testing of drivers that remains the key issue in the campaign to reduce drinking-related traffic injuries and fatalities. Public health experts estimate that one in four motorists drives with BAC levels above the legal limit. In August 1986 the London police offered motorists the opportunity to take a breath test without risking penalty; one out of three of the drivers who volunteered was over the legal limit.[54] According to official estimates, the chances of a driver who is legally impaired getting caught are no more than one in 250, and may be as low as one in a thousand. Only the threat of random testing, according to health promotion advocates, could reduce this threat. Lobbying on behalf of random testing has increased in recent years. Groups including Campaign against Drunk Driving and Action on Drinking and Driving, the Association of Chief Police Officers, and the Royal College of Psychiatrists joined the chorus in support of random testing. The campaign, like its American model, consisted of over 350 parents and other relatives of victims of drunk drivers. In July 1986 it called upon the Thatcher Government to institute random testing.[55] One month earlier the BMA annual conference also came out in favor of the procedure. In addition, there was widespread popular support for random breath tests. A national survey in May 1987 found that 77 percent of the people favored the measure compared to only 25 percent in 1967 and 48 percent in 1975.

The Thatcher government, however, remained opposed to random testing. It believed that it would constitute an unfair and unreasonable infringement of freedom and was guaranteed to strain police-community relations. Critics argued that the real explanation lies in the close political ties between the Tories and the alcohol industry. Derek Rutherford of the Institute of Alcohol Studies told the author that there is "a very powerful influence on this [Conservative] Government" by the alcohol industry in all areas of alcohol control policy. Whatever the reason, and despite increasing expert and lay support for it, the Thatcher government is likely to continue to resist random testing.

[54] *Times* (London), 22 August 1986.

[55] Ibid., 8 July 1986.

Informing the Public: Advertising

The link between alcohol consumption and advertising is at best uncertain. Advertisers and the advertising industry have long objected to regulating alcohol (and tobacco) advertising on the grounds that it is an abridgement of freedom of commercial expression, and that it is unnecessary from the point of view of consumption. Advertising, it is claimed, influences brand choice, not levels of consumption or the initial decision to use a product. Health promotion forces find this assertion intuitively incredible, but they have been frustrated in finding empirical evidence to support their position. Despite a number of studies linking advertising with consumption, the overall evidence is ambiguous.[56] Hence, health promotion advocates typically fall back on commonsensical assertions. Thus, the Royal College of Psychiatrists concluded recently that

> On the whole, there is little convincing evidence at present to refute the alcohol and tobacco industries' oft repeated claim that advertising only affects brand preferences and not overall levels of consumption. As with health education, though, a lack of evidence that advertising increases overall consumption levels is not the same as proof that it does not. It is indeed rather difficult to believe that all those successful young women sipping exotic cocktails, and all those happy and healthy looking workmen downing their pints on so many TV screens and hoardings never persuade someone to have a drink they would not otherwise have had.[57]

In light of both the unproven link between consumption and advertising and the acceptance of the harmlessness of responsible drinking, it is not surprising that there are far fewer limitations placed on alcohol than cigarette advertising. Actually, in Britain there are *no* statutory limitations. Advertising standards have been set by self-imposed codes of practice that restrict certain practices on television and in other forms of advertising. For example, broadcasting advertisements should not include persons who appear younger than twenty-five years old, feature immoderate drinking, link drinking and driving, claim drinking contributes to success, or suggest that solitary drinking is acceptable. As in the United States, there is no television advertising of spirits. It must be emphasized,

[56] One study linking advertising and increased consumption of whisky, but not beer and wine, is McGuiness, *An Econometric Analysis*; for a comprehensive review of the empirical literature on alcohol advertising and consumption see United States Congress, Senate, Subcommittee on Children, Family, Drugs, and Alcoholism, *Alcohol Advertising* hearings (1985), appendix A.

[57] Royal College of Psychiatrists, *Alcohol: Our Favorite Drug* (London: Tavistock Publications, 1986).

however, that these restrictions are far less imposing than those governing cigarettes. There have been no restrictions in Britain on the volume of advertising, the size of advertisements, their location, or their use in sporting events. Furthermore, these codes, one by the Advertising Standards Authority, and the other by the Independent Broadcasting Authority, have not been the subject of formal, periodic government and industry negotiations.

Alcohol advertising in the United States theoretically falls under two general jurisdictions, the FTC and the Bureau of Alcohol, Tobacco, and Firearms (BATF). FTC jurisdiction is limited to claims of deceptive or unfair advertising and promotional or marketing practices. In March 1985 the FTC rejected a petition filed by twenty-nine organizations led by the Center for Science in the Public Interest that charged that "many of the marketing and advertising practices of the alcoholic beverage industry have contributed to the indisputably enormous personal and economic injury connected with alcohol abuse." The petition asked the FTC to ban advertising and promotional and marketing techniques that are aimed at heavy drinkers or young people, use celebrities, associate alcohol with risky activities or success, or encourage excessive consumption. The FTC rejected the assertion that alcohol advertising was either unfair or deceptive, or that it encouraged problem drinking.[58]

The BATF has jurisdiction under the Federal Alcohol Administration Act of 1935 to prohibit "false," "misleading," "obscene," or "indecent" statements in beer, wine, and spirits advertising. The current BATF regulations governing advertising practices date back to the original act and deal essentially with disclosure of ingredients and other representations of alcoholic beverages. In 1978 the BATF issued an Advance Notice of Proposed Rulemaking soliciting comments in preparation for a review of alcohol advertising practices. The result was a rule issued in July 1984 that touched on only a few of the concerns expressed by opponents of alcohol advertising, for example, use of subliminal techniques and therapeutic claims. The bureau indicated that it would continue to consider other issues, including the use of celebrities and athletic events in advertising.

Finally, as in Britain, control of alcohol advertising is in large part left up to trade group and industry self-regulation. The Distilled Spirits Council of the United States, the U.S. Brewers Associations, the Wine Institute, and the National Association of Broadcasters each have their own advertising code or standard. In recent years each has addressed, to some

[58] For a copy of the Federal Trade Commission "Memorandum" see United States Congress, Senate, Subcommitte on Children, Family, Drugs, and Alcoholism, *Alcohol Advertising* hearings, 401–52.

extent, concerns expressed by critics, including those dealing with appeals to women and young people, as well as implied associations between alcohol consumption and success in sports, sexual pursuits, or social achievement. Each group is on record denying that advertising is intended to increase overall or individual consumption of its products.

The empirical evidence and the protestations of the alcohol and advertising industries notwithstanding, there remains considerable public concern over the relationship between advertising and volume of alcohol consumption. This concern, for reasons that should now be obvious, has been articulated longer and more energetically in the United States than in Britain. In particular, two issues stand out: health warnings and a ban on alcohol advertising. In neither country have these issues received as much attention as comparable proposals in the area of smoking. Nevertheless, by the mid-1980s a full-scale national debate had developed in the United States over the issue of banning broadcast advertising, and in 1988 Congress adopted, with remarkably little fanfare, a warning label on alcoholic beverages.

A proposal to ban alcohol advertising on television was first introduced into Congress in 1971, one year after the imposition of the ban on cigarette advertising, lending, by the way, some credence to the "slippery slope" arguments of anti-regulatory forces. Although the bill was stillborn, and the issue lay untouched for a decade—except for Senate hearings in 1976—it was by no means forgotten. In 1981 Congressman George E. Brown, Jr., who would also take the lead in pursuing health warnings on alcoholic beverage products, resurrected the issue by proposing the disallowal of alcohol advertising as a business expense. In addition to raising about $200 million more in federal taxes, this probably would have resulted in an increase in prices and a decrease in consumption. The bill was not, however, reported out of committee.[59]

The most recent, and serious, effort to ban beer and wine advertising from television came in 1985. In that year a coalition of religious groups such as Mormon, Baptist, and Methodist churches, educational groups such as the National PTA, and health advocacy groups such as the National Council on Alcoholism began a nationwide lobbying campaign to convince Congress either to ban wine and beer advertising on television or to allow equal time for counter-advertisements. The coalition was led by the Center for Science in the Public Interest, and was organized around Project Smart—Stop Marketing Alcohol on Radio and Television. Project Smart sought one million signatures to present to Congress. In response to this effort, beverage manufacturers and broadcasters, who earned about $750 million in 1985 from beer and wine advertising, formed their

[59] *Congressional Record* (10 March 1981), 968–69.

own lobbying effort. This included an attempt to create favorable public opinion by starting "the largest and most aggressive campaign that has been launched in the history of broadcasting" to educate viewers on the problems associated with immoderate and irresponsible use of alcohol.[60]

Both sides appeared before Senator Paula Hawkins's (R-Florida) subcommittee on children, family, drugs, and alcoholism, which held a hearing on February 7, 1985, to examine the impact of advertising on consumption. Each side presented by-now familiar arguments. The broadcasters and the alcohol industry contended that (1) advertising simply affects brand preference, not levels of consumption or recruitment of new drinkers; (2) they fully recognized the problem of alcohol abuse and have been acting responsibly to discourage it through education programs; (3) such a ban would be a violation of their First Amendment rights; (4) without advertising certain sporting events such as golf and basketball might disappear from network television; (5) a ban on alcohol advertising could easily lead to banning advertisement of other legal and safe products that, if used improperly, could lead to harm, for instance, automobiles and caffeine and sugar products; and (6) the alcohol industry made a substantial contribution to the economic well-being of the nation in agricultural purchases, packaging, production, sales, promotion, charitable contributions, and taxes. Although no one in the industry mentioned "the goose that lays the golden egg," the sentiment was implicit throughout industry testimony.

For their part, health advocates, too, went over familiar ground. Michael Jacobson, Executive Director of the Center for Science in the Public Interest, rejected most of the assertions of the broadcast and alcohol industries. In particular he charged that advertising seeks to maximize sales, especially among new and young drinkers. In addition, he referred to "credible research demonstrating the effects of advertising on consumption." Finally, he rejected the First Amendment arguments, noting that federal appeals courts have upheld the constitutionality of alcohol advertising bans in both Mississippi and Oklahoma.[61] Although Congress took no action on the proposed ban, the debate over alcohol advertising in the United States will continue. For health advocates advertising, which portrays alcohol as part of a glamorous, rewarding, and socially desirable life-style, runs contrary to the public well-being.

Paralleling the debate over alcohol advertising restrictions has been a personal crusade by Senator Strom Thurmond (R-South Carolina) to require health warning labels on liquor bottles. Senator Thurmond reintroduced the health warning bill, first introduced in 1969, in virtually

[60] *New York Times*, 15 January 1985.

[61] See Senate *Alcohol Advertising* hearings, 243–58.

every subsequent session of Congress. In 1979 the Senate approved an amendment to an alcohol abuse program bill that included a liquor health warning. The health warning provision was dropped in a House-Senate conference committee. Instead, a compromise was adopted requiring the Department of Health, Education, and Welfare and the Treasury Department to conduct a study of the effects of alcohol abuse, especially on infants whose mothers drank during pregnancy. Fetal Alcohol Syndrome is the third leading cause of birth defects. About five thousand infants are born each year with this entirely preventable disease, yet surveys have shown that nearly 60 percent of Americans under forty-five years of age have never heard of the problem. Senator Thurmond hoped the study would lead to recognition of the need for either voluntary or mandatory labeling. It did not.

In 1988 Senator Thurmond reintroduced yet another labeling bill. This time he was joined in the House by the most unlikely of political bedfellows, Congressman John Conyers, a black, liberal Democrat from Detroit. Each man was motivated by a different cause: Thurmond is famous for his dedication to physical fitness and traditional conservative moral values, while Conyers, who represents a predominantly black district, was concerned about the impact of alcohol on black Americans. According to Conyers, "For black Americans, who are the regular and disproportionate targets of alcohol manufacturers, alcohol problems are especially serious: Despite an overall lower rate of drinking in the black community, the diseases are more severe: cirrhosis mortality is twice as common among black men as it is among white men. . . . The rate of cancer of the esophagus, strongly related to alcohol consumption, among black males 35–44 years old is a staggering ten times that of their white counterparts."[62]

Most observers gave the labeling bill little hope of passage. On the Senate side the bill was assigned to the Commerce Committee, which is notoriously pro-business. In addition, according to one congressional aide, "It's always been a burial ground for health bills."[63] The prospects for this particular bill seemed especially bleak, given the liquor-related industry connections of some of the Commerce Committee members. Senators Lloyd Bentsen (D-Texas) and John Danforth (R-Missouri) have both been recipients of campaign contributions from the liquor industry, and Danforth's state is the headquarters for the Anheuser-Busch brewery. Similarly, Kentucky Democrat Wendell Ford, whose state is the largest producer of distilled liquor, and Wisconsin Republican Bob Kasten, whose state is famous for its beer, were also members of the committee.[64]

[62] See *Congressional Record* (9 September 1988), E2887.

[63] Dirk Olin, "This Dud's For You," *New Republic*, 11 July 1988, 12.

[64] Ibid., 13.

The prospects for this perennial tilt at the alcohol windmill did not seem bright. Yet there were other forces at work. First, as in other areas of health promotion, state and local legislative activity was proceeding more rapidly and aggressively. In December 1983 New York City adopted the first law in the nation to require bars, liquor stores, and restaurants to display posters that state, "Warning: Drinking alcoholic beverages during pregnancy can cause birth defects." Since 1986, Georgia, Maine, and South Dakota have required all retail establishments that sell alcoholic beverages consumed on the premises to display a warning sign alerting people to the dangers of drinking liquor during pregnancy. Since 1969 Utah has required state liquor stores and private clubs to post a sign warning that consumption of alcoholic beverages "may be hazardous to your health and the safety of others." Utah's Republican Senator Orrin Hatch has been a supporter of Senator Thurmond's labeling bills for some years. But most significantly for the alcohol industry, in October 1988 California, which is often a state policy trendsetter, became the first state to require health warnings on all alcoholic beverage containers. Since other states had shown interest in similar legislation, alcoholic beverage manufacturers faced the potential prospect of fifty different state warnings.

Second, the alcohol industry feared that it might well face the same rash of product liability claims that the tobacco industry confronted. Indeed, in 1988, several cases were pending against liquor manufacturers, including one in Seattle that was the first to go to trial alleging that a liquor manufacturer (Jim Beam Brands) failed to warn consumers about the link between alcohol use during pregnancy and birth defects. (In May 1989 a federal jury found Jim Beam Brands innocent of negligence for not placing warnings on its beverage bottles.)

Finally, 1988 was an election year in which the issue of drug abuse had percolated to the top of everyone's legislative agenda. No member of Congress could chance being portrayed as "soft on drugs." Supporters of health warnings emphasized to their colleagues and the media that alcohol too was a drug and as potentially dangerous as cocaine or heroin. Thus, whether it was because of its seeming inevitability or its utility, the alcohol industry decided to work out the best deal it could on labeling.

The original Thurmond-Conyers bill, which was endorsed by a broad range of fifty-five educational, health, and civic organizations, would have required five alternating warnings, including one that "alcohol is a drug and may be addictive," and one that alcohol "can increase the risk of developing hypertension, liver disease, and some cancers." In addition, with cigarettes as a model, some members of Congress wanted to include warnings on advertisements. After months of deliberations, a compromise version of the bill was worked out and included in the Omnibus

Antisubstance Abuse Act of 1988. Title VIII amended the Federal Food, Drug, and Cosmetic Act to require a national uniform health warning, beginning in November 1989, on all containers with beverages with at least one-half of one percent alcohol: "Government warning: (1) According to the Surgeon General, women should not drink alcoholic beverages during pregnancy because of the risk of birth defects. (2) Consumption of alcoholic beverages impairs your ability to drive a car or operate machinery, and may cause health problems." In addition to avoiding the more pointed health warnings, the manufacturers won two other concessions: the law specifically preempted all state warning labels, and the warnings did not appear on alcohol advertising. The new law got off to a bumpy start when Senators Thurmond and Gore, as well as Congressman Conyers, complained to the Bureau of Alcohol, Tobacco, and Firearms that under the temporary rules ordered by the agency the new messages were so small and inconspicuously placed as to be unreadable. Senator Gore accused BATF of bowing to alcohol industry pressure and threatened that if they did not change the labels administratively he would introduce legislation to do so. Responding to congressional and health advocacy group demands, in February 1990 the BATF issued final rules specifying the minimum size of the letters for warning labels and requiring that the words "Government Warning" be in bold capital letters. The final rules go into effect in November 1990, although the liquor companies may request a 120-day extension.[65]

Neither a total ban on alcohol advertising nor the imposition of health warnings on beverage bottles has yet attracted much public attention in Britain. Only the BMA, less conservative on health promotion issues than its American counterpart, has been vocal on these two issues, but even it has been somewhat tentative and inconsistent. At its annual meeting in June 1985 the BMA approved a motion calling for a complete ban on the advertising and promotion of alcohol. Unlike other life-style related measures approved at the meeting, including compulsory fitting of rear seat belts, a ban on tobacco advertising and promotion, and tougher penalties for shopkeepers who sell cigarettes to children, the motion on the alcohol advertising and promotion ban was highly controversial and was approved by only a narrow margin. Opponents of the motion reminded their colleagues that it was the misuse of alcohol, not responsible use, that was the problem.[66]

Opposition to a proposed ban persisted within the BMA, and in the following year the association voted to rescind its call for a ban on advertising and promotion. It called instead for more "sensible drinking" and

[65] *New York Times*, 11 February 1990.

[66] *Times* (London), 27 June 1985.

health warnings on alcoholic beverage packages and advertisements as part of "an effective and sustained [Government] campaign aimed at reducing alcohol-related problems." One of the delegates to the 1986 annual meeting explained his support for the shift in the following way, "If we continue with a policy as extreme as the one we have then the association is less likely to be taken seriously by government, who will see it as impractical; by the drinks industry, who will see neither room nor encouragement for compromise and progress; or by the general public who may perceive it as hypocritical."[67]

Recently there has been growing government concern and public hostility toward acts of violence involving young people apparently under the influence of alcohol, especially at soccer matches both in Britain and abroad. A new term, "lager louts," has entered the vernacular to describe drunken youths who engage in various acts of hooliganism. This concern reached fever pitch during 1988 and resulted in a variety of proposals to curb public drunkenness, including strengthening of laws to prevent underage drinking, more aggressive enforcement of existing public disorder laws, introduction of identity cards, restriction of the sale of alcoholic beverages around soccer fields, tax benefits to producers of low alcohol drinks, more rapid judicial processing of public disorder cases, and changes in advertising. A parliamentary committee looking into the problem of alcohol abuse reported that one-half of fifteen-year-olds drank alcohol on a regular basis. The chairperson of the committee, Lady Masham, recommended to the government that alcohol advertising be banned in the movies and on television. Her recommendation went unheeded.[68]

Although the concern with public drunkenness has taken on an unusual sense of urgency, the historical British attachment to social drinking and the general but by no means uniform reluctance of the Thatcher government to regulate either business or social behavior mean that proposals to ban alcohol beverage advertising in the near future or to require warning labels face extraordinary obstacles. The latter measure would, of course, not be inconsistent with Thatcher's penchant for limiting government's role to encouraging informed citizen choice. Nevertheless, given the British tradition of voluntary action rather than public regulation, the absence of neotemperance sentiment, the economic and political power of the alcohol beverage industry, and the continued importance of alcohol in British social life, it will not be easy to gain approval for either measure. Continued youthful alcohol-associated violence may, however, lead to some limitations on advertising aimed at young people.

Finally, it should be noted that the alcohol industry has responded in

[67] *Times* (London), 25 June 1985.

[68] *Economist*, 25 June 1988, 62.

much the same manner as the tobacco industry to the increased economic participation and influence of women. Recognizing that women represented an opportunity for expanding sales, particularly at a time of declining male demand, alcohol companies have begun targeting them. In recent years both Brown-Forman, makers of Jack Daniels, and Seagrams have geared campaigns and new products toward women. In Britain, corporate efforts to increase sales among women have involved attracting more women into the pubs, which are often tied to specific breweries, through enticements such as women's happy hours and gifts.[69]

Alcohol use is a far more complicated issue in the life-style and public health policy debate than tobacco use. This is reflected in the fact that fewer of the questions raised in the evaluative framework lend themselves to readily acceptable answers. There has been increased acceptance of the notion that some forms of alcohol abuse have biological origins, but there is by no means agreement over how much of the problem can be justified in this regard. There is, to be sure, renewed concern about the implications of the biological model, with some health advocates arguing that it either has been used to excuse personal irresponsibility or to divert policy attention from the majority of those who misuse alcohol through irresponsible behavior.

In addition, the line between harmless and harmful behavior is not always either scientifically or behaviorally clear, as is the case with smoking. It has been difficult to fashion politically acceptable public policies that can at the same time deter abuse yet not penalize "responsible" consumption. Where a clear and provable link does exist between life-style choice and social harm, as with drinking and driving, or violence at sporting events, there has been willingness in both countries to take strong policy action.

The shared context of medical and social complexity and uncertainty concerning the use of alcohol is mediated through the different cultural norms that prevail in each country. The result is very different British and American approaches to alcohol control. In Britain, where cultural values play a minimally inhibiting role in the use of alcohol, public policymakers have shown little inclination to restrict accessibility and availability of alcoholic beverages as a way of reducing consumption and harm. In recent years the British have rejected random testing, extended pub hours and places where alcoholic beverages can be sold, and implicitly rejected increasing the legal drinking age. The British have, however, been much

[69] Cavanagh and Clairmonte, *Alcoholic Beverages*, 133–35.

more willing to increase the cost of drinks, but not at a rate that significantly increases the real price of the beverage. Furthermore, in the election year 1987, the Thatcher government was unwilling to impose even this inhibition. One would not expect greater decentralization of political power to lead to more restrictive policies on alcohol. The major obstacle to more aggressive alcohol control in Britain lies not in the wealth and power of the drinks industry—although it is considerable—but in cultural values and social preferences. Finally, it should be recalled that Thatcher's preference for removing as many restrictions as possible from the economic behavior of both consumers and providers of goods and services has further facilitated an essentially laissez-faire approach in this policy area.

In the United States social norms, too, have played a key role in defining policies dealing with alcohol use, sales, and promotion. Neotemperance and selective prohibitionist sentiment is strong, at least among the health-conscious middle and upper classes, as well as among political, social, and religious conservatives. Specific situations, including the lobbying effort by MADD and the 1984 and 1988 presidential elections, provided a stimulus and an occasion for the adoption of key alcohol control policies: a national minimum drinking-age law and a health warning law. Both of these measures were adopted despite considerable political and economic obstacles.

Another structural factor influencing American policy was the impact of federalism. As in the case of cigarettes, states and localities acted earlier than the federal government to recognize and respond to the social harm caused by alcohol abuse. The prospect of fifty different, and perhaps more alarming, warning labels undoubtedly affected the alcohol industry's decision to accept a labeling law. With regard to the drinking-age law, the force of public opinion and the threat of loss of federal highway funds were sufficient to overcome the reluctance of some states to adopt the national standard. It is important to remember that this latter law, the most restrictive of the actions taken in response to problems growing out of alcohol abuse, and the one in which the tension between personal freedom and public health was most acute, involved a group in American society who stood just on the threshold of political adulthood. Charges of paternalism, and even Big Brotherism, were less damning in this instance than they might otherwise have been.

Six

Road Safety Policy: Blaming The Car Or The Driver?

SINCE the first recorded traffic death in 1899, nearly 2.5 million Americans have died in road accidents, more than twice the number killed in all the wars in our history. In Britain the story is much the same. In a 1979 parliamentary debate on mandatory seat-belt legislation, one member of Parliament pointed out that in the previous five year period (1974–1979) there had been over three times as many casualties on British roads as British troops suffered during the five years of World War II.[1]

In recent years policymakers in both the United States and Britain increasingly have concerned themselves with the "carnage on our highways." The irony of this increased interest is that it began at a time when road deaths were declining in both countries. Between 1971 and 1982 (the last year before implementation of a national seat-belt law), road deaths in Britain fell from 7,970 to 6,124, despite an increase in the number of vehicles and vehicle miles traveled. Similarly, in the United States, the death rate from motor-vehicle accidents declined from 27.4 per 100,000 in 1970 to 19.2 per 100,000 in 1986. In both countries the decline in fatalities and serious injuries was attributable to a combination of factors, including improved automobile safety designs and lowered speed limits following the 1973–1974 Arab oil embargo.

Despite the overall progress in reducing highway death rates over the past decade, automobile safety became the focus of increased attention in both countries in the 1970s and 1980s. The most obvious reason for this is that automobile accidents were transformed into an almost exclusively life-style-related problem; many, especially those with conservative political inclinations, attributed the problem largely to unsafe driving habits rather than, say, unsafe roads or automobiles making it a matter of personal rather than corporate or government responsibility. A second explanation for this increased concern is that the rate of decline in road accidents has lagged behind accident-rate declines in the home and workplace. Between 1961 and 1983 workplace accidental deaths in Britain fell 65 percent, and deaths from railway accidents fell 62 percent, while traffic fatalities declined only 22 percent. Furthermore, "compared with in-

[1] *United Kingdom, Parliamentary Debates* (Commons) (March 1979), col. 1779–80.

dustrial accidents, road accidents have received a far smaller proportion of administrative resources devoted to accident prevention."[2] In the United States, while the overall accidental death rate declined 29 percent from 1960 to 1984, the motor-vehicle death rate declined only 15 percent. A third reason for policy concern is that young people account for a disproportionate share of serious injuries and fatalities. In the United States, the 1986 overall motor-vehicle accident rate was 19.1 per 100,000, but the rate for fifteen-to-twenty-four-year-olds was 36.5 per 100,000. This latter figure was higher than the 1950 death rate for the same age group, the only one for which there was no decline in deaths over the last three and one-half decades. Although comparable figures were not available for Britain, the most recent data indicate that fifteen-to-twenty-nine-year-olds are three to four times more likely to suffer serious traffic injuries or deaths than other age groups.

The impact of the youthful nature of this health problem on policymakers has been obvious. In her statement announcing the Transportation Department's 1984 decision to require the installation of air bags by 1989 unless two-thirds of the U.S. population were covered by mandatory seat-belt laws, Transportation Secretary Elizabeth Dole said, "In the last 10 years, 470,000 Americans died on our highways. And sadly, most of the victims are young people, young people whose lives could be saved."[3] Similar concern was expressed recently by a district medical officer in his testimony before the Transport Committee of the House of Commons, which was conducting an inquiry into road safety. Bemoaning the small amount of money spent on road safety research, he said, "Contrast the enormous amount of money put into research into heart disease and cancer which tend to afflict older people, with the paucity of money spent on research into accidents which affect mainly young people with a much greater 'life potential.' "[4] The tragic association between young persons and automobile fatalities and injuries has made road safety a major health promotion issue.

Finally, increased attention to automobile safety must be viewed in the context of the overall new-perspective concern with preventable diseases and, especially, their economic consequences. It is difficult to place a precise price tag on the costs of traffic deaths and injuries to society, but both American and British policymakers have estimated these costs to be considerable. In the United States, the National Highway Traffic Safety Administration calculated that, in 1980 dollars, the costs to society of motor-vehicle accidents in medical care, lost productivity, and insurance

[2] United Kingdom Parliament Transport Committee, "Second Special Report from the Transport Committee: Road Safety" (Commons) (1983), 134.

[3] *New York Times*, 12 July 1984.

[4] United Kingdom Parliament Transport Committee, "Second Special Report," 93.

administration was $57 billion ($234 per capita).[5] In Britain road accidents cost society about £5 billion annually. In both countries policymakers are convinced that road accidents are a significant factor in the high and rising costs of health care and these accidents are, to a considerable extent, the result of personal recklessness.

Despite the concern over road safety in both countries, the problem is perceived to be of less urgency and importance in Britain than in the United States. The notion that the British are somewhat "indifferent" to the problem is a conclusion reached in the most recent British government study on road safety, the 1987 *Interdepartmental Review of Road Policy*. Among the major conclusions of the *Review* is that "The subject of road safety appears, by and large, not to be regarded within 'opinion forming' circles—politicians, political observers, the pressure groups, the media—as a particularly interesting or important one."[6] The *Review* offered as evidence in support of this conclusion the fact that there had not been a government White Paper—an obtrusive measure of policy concern—on road policy in two decades, and that a 1985 Commons debate was the first one on the subject in living memory. (This neglect was remedied in February 1989 with the publication of the White Paper *The Road User and the Law: The Government's Proposals for Reform of Road Traffic Law*.) In addition, party manifestos routinely ignore road safety, and there are no automobile safety interest groups comparable in organization, sophistication, and leadership to those in other public health areas. Finally, there appeared to be "a lack of sustained interest in road safety on the part of the public at large, or at any rate . . . the absence of any obvious manifestation of serious public concern about the issue."[7]

I am not suggesting that road safety has been ignored by British policymakers, only that it has attracted less attention than some other public health and safety issues. The Working Group attributed the relative lack of interest to several factors including a failure on the part of the political elite to appreciate the human and material magnitude of the problem, the fact that road accidents attracted little media attention, and "an attitude of sheer fatalism—a feeling that accidents on the road are a long-standing and inescapable feature of a motorised society and that we have simply no alternative but to continue to put up with the problem."[8]

From an Anglo-American comparative perspective there is probably another reason for the relative "indifference" to road safety. Compared to

[5] Robert W. Crandall et al., *Studies in the Regulation of Economic Activity* (Washington, D.C.: Brookings Institution, 1986), 45–84.

[6] United Kingdom Department of Transport, *Interdepartmental Review of Road Policy Safety* (London: Her Majesty's Stationery Office, 1987), 19.

[7] Ibid., 19.

[8] Ibid., 19–20.

accidents in the United States, and compared to other public health problems, road accidents in Britain *are* less significant in terms of lives lost and money spent. In the United States the automobile death rate in 1986 was 19.1 per 100,000 people compared to 9.2 per 100,000 in Britain. The comparative per capita cost to society of road deaths and injuries was $246 per person in the United States and $73 per person in Britain. That such a difference should exist is perhaps predictable given the fact that there are more than twice as many private cars per one hundred people in the United States (fifty-six) than in Britain (twenty-nine); about 10 percent of the households in the United States did not own a car in 1986, compared to nearly 40 percent in Britain.

Finally, although the 1989 White Paper on traffic law reform signals increased British interest in this area, it is unlikely that there will be much of a change in the relative neglect under the Thatcher government. When the Interdepartmental Working Group was set up by the government to examine road safety it was instructed to keep the following considerations in mind: "No increase in overall resources available for road safety should be assumed," and "In examining future options there should be a presumption against measures which involved the imposition of new legislative controls on road users, except where unavoidable."[9] A Government that does not intend to spend more money on road safety and is reluctant to impose regulations to improve driving habits is limited in what it can accomplish.

After first reviewing some general issues and options in road safety policy in each country, I will devote most of the chapter to one of the more prominent life-style and health debates of the last decade, that is, should government mandate the use of seat belts? This issue, more than any other road safety measure, reflects the tension between personal freedom and public health. It is a protypically new-perspective issue.

Automobile Safety Policy Options

Before turning to the specific policy responses of Britain and the United States to the problem of motor-vehicle deaths and injuries, it is useful to remember that motor-vehicle accidents result from one or more of the following factors: road conditions, including lighting, markings, maintenance, road engineering, and specific surface conditions; traffic rules, management, and enforcement; vehicle safety equipment and construction; and drivers, or the "human factor."

Since this book is concerned with the impact of personal behavior on

[9] Ibid., 3.

health, the emphasis in this chapter is on the human factor in the road safety equation. This emphasis is justified for two reasons. First, human factors play the predominant role among the four factors in motor-vehicle accidents in both countries. The 1987 report by the Inter Departmental Review of Road Safety Policy in Britain concluded that, "Human error is the prime factor in 70% of accidents and one factor in 95% of accidents."[10] Whether it is through carelessness, inattentiveness, recklessness, ineptness, or driving under the influence of drugs or alcohol, most road accidents are attributable to the driver rather than the vehicle or road conditions. It must be emphasized that poorly designed cars or roads also contribute to the nature or likelihood of injuries or deaths. However, there is both policy and academic emphasis on the human factor in road safety. Second, the emphasis on human factors reflects the policy shift in both countries from emphasizing vehicle design and road conditions to making individuals more accountable for their behavior.

The United States: Safe Cars versus Safe Drivers

During the early 1960s there was an increase in the U.S. motor-vehicle death rate, from 20.8 deaths per 100,000 in 1961 to 25.3 per 100,000 in 1965, after nearly two decades of decline. Perhaps more politically important, the absolute death rate was rapidly approaching a critical symbolic threshold of 50,000 deaths per year. In 1965, 49,000 Americans lost their lives in traffic accidents; the following year the number reached 53,041. With the number of automobiles rapidly increasing in affluent America, from 49.2 million in 1950 to 91.3 million in 1965, the prospects for continued devastation on the highways seemed assured.[11]

The concern with motor-vehicle accidents and safety received additional impetus with the 1965 publication of Ralph Nader's expose of the safety record of the automobile industry in general, and General Motors' "Corvair" in particular. In addition to documenting that the "Corvair" was an unsafe automobile, *and* that General Motors knew it to be so, Nader alleged, in *Unsafe at Any Speed*, that the industry generally disregarded safety in its preoccupation with profit. Furthermore, he suggested that although it was true that drivers were often responsible for the "first collision," much of the human and material damage occurred in the "second collision," when drivers hit unsafe windshields, dashboards, and steering columns. Nader claimed that, contrary to the industry's position,

[10] Ibid., 135.

[11] United States Bureau of the Census, *Statistical Abstract of the United States, 1987* (Washington, D.C.: Government Printing Office, 1986), 586.

little could be done about improving drivers' behavior, but a great deal could be done about improving the safety of motor vehicles.

> Furthermore, our society knows a good deal more about building safer machines than it does about getting people to behave safely in an almost infinite variety of driving situations that are overburdening the driver's perceptual and motor capacities. In the twenty to forty million accidents a year, only a crashworthy vehicle can minimize the effects of the second collision. Vehicle deficiencies are more important to correct than human inadequacies simply because they are easier to analyze and to remedy. And whether motorists are momentarily careless or intoxicated, or are driving normally when they are struck by another vehicle is entirely irrelevant to the responsibility of the automobile makers to build safer cars.[12]

The impact and credibility of Nader's study was enhanced when the president of General Motors admitted before a Congressional committee that the company had hired private investigators to follow Nader. The notoriety attending the book, and the General Motors revelation, furthered the cause of automobile safety reform. On February 2, 1966, President Lyndon Johnson referred to "the slaughter on our highways" as the "gravest problem before this nation—next to the war in Vietnam."[13] Johnson submitted automobile and highway safety bills to Congress in March. In the summer of 1966, following hearings and overwhelming Congressional support, two bills were approved unanimously: the Highway Safety Act, and the National Traffic and Motor Vehicle Safety Act.

THE HIGHWAY SAFETY ACT

The highway act was "designed to provide federal guidance to the states for highway construction, drivers' education, traffic law enforcement and other highway safety programs."[14] It required each state to establish a highway safety program consistent with federal standards set by a newly created Highway Safety Administration, and provided matching federal funds to implement the program. The law stipulated that a state could lose 10 percent of its federal highway construction funds, and would not receive any highway safety funds, if it failed to adopt an approved program.

Motorcycle Helmets

It is somewhat ironic that one of the first standards issued under the Highway Safety Act dealt not with road conditions or other engineering prob-

[12] Ralph Nader, *Unsafe at Any Speed: The Designed-In Dangers of the American Automobile* (New York: Grossman Publishers, 1965), 344.

[13] Quoted in Congressional Quarterly, *Congress and the Nation* 2 (1966): 785.

[14] Crandall et al., *Studies*, 46.

lems, but with motorist behavior. It was announced in 1967 that the states should adopt laws that drivers and passengers of motorcycles wear approved safety helmets; failure to do so would result in a loss of federal highway construction and safety program funds. The secretary of transportation justified this standard by pointing out that deaths and injuries from motorcycle accidents had doubled between 1963 and 1965, that it was a problem particularly afflicting the young, and that given the increasing number of motorcycles the tragedy was certain to get worse.[15]

The highway safety standard had an immediate impact on the states: the number of states without mandatory helmet laws rose from zero in 1965 to forty in 1969 and forty-seven in 1975.[16] As one might expect, the laws were opposed by motorcycle enthusiasts, including the American Motorcycle Association, and political leaders who saw this as both an intrusion on individual freedom and a case of federal Big Brotherism. The helmet laws were challenged in more than one dozen states, with some success.[17] Central to these challenges was an issue that goes to the heart of the debate over public efforts to modify risky behavior. This issue was framed in the following manner in *State of North Dakota v. Odegaard* (1969), in which the defendant appealed a conviction for failure to wear a helmet. "It is Mr. Odegaard's contention that the issue resolves itself into this question: 'Is it constitutionally permissible for the state to abridge the liberty of an individual by compelling him, contrary to his desires, to adhere to a course of conduct ostensibly calculated to protect him from possible harm when such conduct has no relationship to the health, safety and welfare of other persons or the public at large?' "[18]

In short, Odegaard's contention was that failure to wear a helmet was a self-regarding activity, endangering no one but himself and that, therefore, the state was unjustified in imposing its will on motorcyclists. The North Dakota Supreme Court disagreed. It accepted arguments from cases in Rhode Island and Massachusetts and found that a motorcyclist who suffers head injuries because of failure to use a helmet does in fact impose costs, or negative externalities, if not necessarily physical harm, on others. This position was endorsed in a Federal District Court decision upholding the Massachusetts law:

> We cannot agree that the consequences of such injuries are limited to the individual who sustains the injury. . . . From the moment of the injury, society picks the person up off the highway; delivers him to a municipal hospital and municipal doctors; provides him with unemployment compensation if, after recovery,

[15] Quoted in Kenneth Royalty, "Motorcycle Helmets and the Constitutionality of Self-Protective Legislation," *Ohio State Law Journal* 30 (1969), 358.

[16] "Motorcycles, Safety, and Freedom," *Regulation* (July–August 1980): 10.

[17] For a review of these cases see Royalty, "Motorcycle Helmets," 355–81.

[18] *State v. Odegaard*, 165 N.W. 2d 679.

he cannot replace his lost job, and, if the injury causes permanent disability, may assume the responsibility for his and his family's subsistence. We do not understand a state of mind that permits plaintiff to think that only he himself is concerned.[19]

The ultimate fate of motorcycle helmet laws, however, was decided not by state or federal courts but by the U.S. Congress. It happened in the following way. In 1975 the National Highway Traffic Safety Administration (NHTSA) decided to bring administrative action, as provided by the 1966 Highway Safety Act, against the three states (California, Utah, and Illinois) that had not yet adopted mandatory helmet laws. Opposition from the congressional delegations of these states, and continuing lobbying efforts by the American Motorcycle Association, finally convinced Congress in March 1976 to rescind the NHSTA's authority to withhold highway funds from those states not complying with the helmet standard.[20] With this decision the legislative rush moved in the opposite direction. By 1988 only twenty-one states still had helmet laws that applied to all riders, although twenty-five required helmets of riders under eighteen years of age. The debate, however, is not over. In 1988 Oregon and Nebraska resurrected helmet-use laws for motorcycle drivers and passengers regardless of age, and other states, faced with mounting health care costs, were considering them as well.

THE NATIONAL TRAFFIC AND MOTOR VEHICLE SAFETY ACT

The purpose of the National Traffic and Motor Vehicle Safety Act was "to reduce accidents involving motor vehicles and to reduce the deaths and injuries occurring in such accidents." Jurisdiction over the law was first granted to the Secretary of Commerce. In 1967, as another measure of the importance attached to road safety and other transportation issues, the Department of Transportation was created, and its secretary became responsible for administering the act. Under the law the secretary was empowered to establish safety standards for motor vehicles to protect the public "against unreasonable risk of death and injury in motor vehicle operation." The act created a National Traffic Safety Agency that, in April 1967, was consolidated with the National Highway Safety Agency of the new Department of Transportation into the National Highway Safety Bureau (NHSB). (To complete the bureaucratic odyssey of road safety administration, in 1970 the NHSB became the NHTSA).

Since its inception the safety agency has been authorized to establish national vehicle safety standards to improve accident avoidance, crash

[19] Quoted in *New York Times*, 29 December 1975.

[20] "Motorcycles, Safety, and Freedom," 10.

protection, and occupant survivability. During its most active years, between 1967 and 1974, the NHTSA issued forty-five mandatory federal motor-vehicle safety standards covering such safety features as tires, brake systems, windshields, defrosting and defogging devices, head restraints, mirrors, occupant and exterior body protection, fire retardation, and steering systems. The automobile industry opposed almost all of the proposed standards and was able to win modification or delays in implementation of many of them. Strong support from legislators and public safety interest groups helped the NHTSA to successfully pursue its mandate during its first years. The agency had, in effect, adopted the Nader position that the greatest room for improvement in road safety lay in modifying automobile design, rather than driver behavior. However, in an important measure of the change in ideology, under the Reagan administration the NHTSA issued just one new vehicle safety rule: the requirement of a third, high-mounted rear brake light on new automobiles. Consistent with conservative ideology, primary responsibility for road safety was again shifted to driver behavior.

One of the first safety features ordered by the agency was seat belts. It required that all new cars sold in the United States after March 1, 1967, be fitted with seat belts. Manufacturers had begun installing front lap seat belts in 1964 largely because more than one dozen states required them to do so. The new federal standard made installation of front lap and shoulder belts, and rear lap belts, mandatory for all automobiles. Unfortunately, while the evidence was overwhelming that seat belts reduced the risks of death or serious injury by at least 50 percent, acceptance of seat-belt use by drivers and passengers was not. Studies indicated that no more than 25 percent and as few as 10 percent of automobile occupants used the restraints.

In an effort to circumvent motorist apathy (or irresponsibility), in the late 1960s the NHTSA began examining "passive restraint systems." In 1969 the NHSB ordered the installation of an "Inflatable Occupant Restraint System," or an air bag, in all new cars as of January 1, 1972. Thus began one of Washington's longest running policy shows. I will take up the story of restraint systems later in this chapter.

Britain: "Human Error, Folly, and Miscalculation"

In a June 1960 debate on road safety, a member of Parliament said that "road accidents [are] the scourge of this century, just as tuberculosis and squalor and ill health were of previous generations." The legislator called upon the government to "tackle" the problem. In his response, the Conservative minister of transport acknowledged the seriousness of the problem but was quoted as saying, "In every accident resulting in a killing he asked himself what he, as Minister could have done to prevent it, and the depressing answer was that in the majority of cases no Minister could have done anything, the death being the result of *human error, human folly, or human miscalculation*."[21]

The prevailing wisdom in Britain, as in the United States, was that the primary culprit in automobile injuries and fatalities was the driver, not the vehicle. Contrary to the situation in the United States, where this view was tempered, although not entirely replaced, with the passage of the 1966 automobile and highway safety legislation, in Britain it prevailed through both Labor and Conservative governments. The British have been remarkably slow to adopt many proven automobile safety devices. For example, despite lobbying efforts by health and safety experts, the British automobile industry still has not been required to install laminated windshields, which have been proven effective in reducing eye injuries and facial lacerations. Such windshields are required in vehicles in the United States, Canada, and most parts of Western Europe. Similarly, it was not until 1986, twenty years after the United States had done so, that the British required the installation of rear seat belts; as of 1989 they still did not require head restraints, or anti-burst door latches. Furthermore, only recently have they discussed legislation that would require minimum safety standards to protect occupants from side-impact accidents. Such standards exist in the United States, but not in Europe. The relative lack of progress on issues of vehicle safety is, I think, a reflection of both the lower priority assigned this policy area in Britain compared to the United States, and the fact that in terms of automobile design, Europe, not the United States, is the more "significant other" for British policy.

The major source of policy conflict in this as in other areas of public health over the past quarter century has been over how to reconcile the competing demands of public health and safety and individual freedom. This conflict is reflected in debates over certain key traffic safety policies. The discussion that follows focuses briefly on two of these issues—establishing speed limits and requiring the use of motorcycle helmets—while a third, controlling drinking and driving, was discussed in Chapter 5.

SPEED LIMITS

In 1977 the Labor minister of transport, under considerable pressure from motorist groups and members of Parliament of both parties, announced that the decreased speed limits introduced during the 1973–1974 oil crisis would be rescinded. The minister acknowledged that rais-

[21] *Times* (London), 2 June 1960.

ing speed limits might well lead to increased road injuries and fatalities. The dilemma for the policymaker, however, was that "we want to save life, but we like driving fast."[22] Since the first speed limits were established in Britain in 1934, the conflict between the freedom of each individual motorist to determine a safe speed and the evidence that controlled speeds generally lead to reductions in road accidents has caused considerable policy conflict.

The issue became most contentious, however, following the imposition of reduced speed limits on motorways and carriageways (that is, highways and freeways) in December 1973, following the Arab oil boycott. Speed limits were reduced to fifty or sixty miles per hour, depending upon the road, to conserve fuel. Between December 1973 and April 1977 the government experimented with various combinations of limits as popular sentiment waxed and waned over the restrictions. The impetus for the change was energy conservation, but an attending consequence was a 14 percent decline in traffic injuries and fatalities between 1973 and 1975. After the energy crisis "disappeared," public health officials and medical interest groups urged the Labor government to maintain the reduced limits. Motoring organizations, especially the highly influential Royal Automobile Club, lobbied against the limits. In April 1976 the Conservative party spokesman for transport issues, Norman Fowler, began aggressively attacking the limits. He believed that from an energy perspective they were no longer needed, and that they were creating tension between the police and motorists.

In October 1976 the Labor government extended the speed limits for six months and announced that it would solicit the opinions of both the general public and specific organizations. In April 1977, having consulted with over fifty organizations—although not any medical associations—the minister for transport announced that speed limits would be raised to sixty and seventy miles per hour on British highways. The minister acknowledged the life-saving consequences of reduced speed limits but also recognized public pressure to "drive fast." With regard to faster speed limits he said, "[They are] dangerous in some respects, but that is life."[23] People, in short, must be free to choose even foolish behavior.

MOTORCYCLE HELMETS

The British Department of Transport has estimated that "for every mile traveled the motorcyclist is 20 times more likely to be killed or seriously

[22] Quoted in "Speed Limits and the Public Health," *British Medical Journal* no. 6068 (23 April 1977): 1045.

[23] Quoted in ibid.

injured than the car driver."[24] These deaths occur primarily among young males below the age of twenty-five, thus magnifying the sense of tragedy. The problem seemed particularly urgent in the early 1960s following dramatic increases in motorcycle deaths and serious injuries. Between 1954 and 1961 there was a 37.9 percent increase in deaths, from 1,148 to 1,544, and a 64.6 percent increase in serious injuries, from 15,847 to 26,085. The Conservative government recognized that some action was necessary to meet this problem but faced a dilemma with which it would grapple many times, in many other health promotion areas, over the next several years. Historically the Conservative party has managed to accommodate both the libertarian and paternalistic traditions of the British political culture in its programs and party philosophy. It is only recently, under Thatcher, that it has virtually jettisoned paternalism. In any event, in 1962 the minister had to reconcile both inclinations in his proposal for reducing motorcycle injuries and fatalities. His solution was a contradictory one.

First, he outlined the reasons why mandating the use of protective headgear was the most effective solution to the problem. Cyclists were vulnerable to injuries, especially head injuries. Research demonstrated that the use of helmets could reduce those injuries by 30 to 40 percent. Motorcycle accidents produced negative externalities and were not simply self-regarding episodes—"a large amount of medical attention, which has to be paid for, is given to him which might be better devoted to some other people." And educational campaigns had only been partially successful in converting cyclists to helmet use; between 1956 and 1962 the usage rate increased from 40 to 66 percent.

Having established the case for requiring the use of helmets, the minister went on to say that he was now only asking for the authority to mandate their use, but would not do so immediately. Instead he hoped to persuade motorcyclists of the foolishness of not wearing helmets. "If by propaganda we are successful, there will be no need for the regulations. On the other hand, if the persuasive quality of the Minister of the day . . . fails, the regulations will have to be brought in, because we want to protect the motorcyclists, in spite of their stubbornness in wishing not to wear helmets."[25]

Because the bill was sponsored by the government it was handily approved. However, it would be over a decade before a minister for transport would use the authority to mandate use of motorcycle helmets. It was not until February 1973, when the Conservatives were back in power, that the minister for transport announced that beginning June 1,

[24] United Kingdom Parliament Transport Committee, "Road Safety,": (Commons) 132.

[25] *United Kingdom Parliamentary Debates* (Commons) (17 July 1962), cols. 256–57.

1973 motorcyclists would be required to wear helmets. As the minister explained it, despite a decade of exhortation, only about 75 percent of all motorcyclists—compared to 66 percent in 1962—used helmets. Since the authority to issue the regulation already existed, opponents could only register their disapproval. They were able to do this formally in a "take note" motion, a parliamentary device by which the House recognizes an executive decision and has an opportunity to comment on it. The debate provides a classic summary of the concerns over life-style issues.

First, opponents argued that the state should not become involved in regulating self-regarding behavior. "There cannot be a free society if its law-making body interferes with people on occasions when they can damage only themselves and not other people."[26] Second, they argued that once one begins regulating behavior that may have some indirect consequences, such as the mental anguish of parents who have lost a child in an accident, there is no end to the mischief in which government might become involved. "If we do this thing on such grounds, we shall be laying the basis for a series of new laws which will reach right into every act, every form of behavior, every choice of the average citizen."[27] Third, unpopular or unreasonable laws such as this one tend to have a negative effect on law in general. "The law falls into disrespect if it tries to lay down conditions which ordinary people will not accept." The debate, while frequently sharp, merely provided opponents an opportunity for catharsis. Beginning in June 1973, British motorcyclists were required to wear approved helmets. The public debates over helmet use in both countries adumbrated, but in a comparatively modest scale, the policy battle that was shaping up over mandating seat-belt use.

Saving Lives and Protecting Liberty: The Seat-Belt Debate

The debates over mandatory seat-belt laws in Britain and the United States have raised some of the most fundamental issues arising out of the new-perspective emphasis on modifying putatively risky personal behavior. Proposed seat-belt laws generated considerably more public passion, however, because in addition to limiting freedom, they also affect many more lives, given the central role the automobile plays in postindustrial societies such as the United States and Great Britain. (In 1987 there were 5.1 million motorcycles compared to 139 million passenger cars in the United States.) For all practical purposes the motoring public and general

[26] United Kingdom Parliamentary Debates (Commons) (5 April 1973), col. 748.
[27] Ibid., col. 760.

public are, in the words of the London *Times*, "more or less coextensive."[28]

In the remainder of this chapter I will examine and compare the debate over seat-belt legislation in both countries. This task was relatively easy for Britain since Parliament sets national law in this area. It was more difficult in the United States since it is impossible to examine the debates that have taken place in nearly all fifty states. I have chosen, then, three states—New York, Illinois, and Oregon—for case studies, although references will be made to others. The three states were chosen, in part, because of availability and accessibility of information, and particularly because they record legislative debates, something which not all states do. New York was also chosen because it was the first state to adopt a mandatory seat-belt law, and because its legislature did so before the 1984 Department of Transportation decision on air bags and passive restraints. Illinois, whose economy supplies a great deal of the material and accessories for the automobile industry, adopted its law after the decision. Oregon, an agricultural state, was chosen because its legislature initially rejected a seat-belt bill in 1985 and then two years later approved a one-year trial period for a seat-belt law, after which time the issue would be referred to the voters. Ultimately the voters rejected the measure through an initiative even before the trial period could begin, although as of this writing yet another attempt to bring the issue to the voters has been started. The three states represent an economically, geographically, and politically diverse, if not statistically significant, sample.

Great Britain: "The Greatest Epidemic of Our Time"

Although parliamentary questions about mandating seat-belt use were raised as early as 1959—and front seat belts had been required equipment since 1965—the first formal proposal to mandate their use occurred in 1973. The proposal was introduced into the House of Lords as an amendment to a Road Traffic Bill but was lost with the dissolution of Parliament in February 1974. Over the next seven years there were six additional proposals to adopt a seat-belt law. Most of these efforts took the form of private member's bills, although on one occasion the Labor government introduced the measure. On three occasions (March 1976, March 1979, and July 1979) seat-belt bills actually were approved by comfortable margins on a second reading in the House of Commons. Each time, however, a general election or the end of a parliamentary session intervened to prevent the measures from completing the legislative process.

[28] *Times* (London), 2 April 1962.

In December 1980, for the seventh time in as many years, a bill to require the use of seat belts by front seat occupants was introduced as a private member's bill in the House of Lords. It was approved (72 to 36) on second reading, after a five-hour debate. Private member's bills are notoriously difficult to get enacted because of time limits set on debate. As Lord Nugent, the bill's sponsor, explained, "I did not proceed with my Private Member's Bill because of the well-known vulnerability of Private Member's Bills to any opposition in the House of Commons. One determined opponent is quite enough to destroy a bill."[29] Lord Nugent was intent on finding a better legislative vehicle for it.

The opportunity arose within a few months. In May 1981 the government introduced a major transport bill. One part of the bill required the use of seat belts by children up to the age of thirteen if sitting in the front seat of an automobile. Allying himself with pro-seat-belt members in the House of Commons, Lord Nugent sought, unsuccessfully, to have the Transport Bill amended to include a mandatory seat-belt-use provision. Opponents of the measure, who included the secretary of state for transport, were able to block it. The Transport Bill left the House of Commons without any action taken on compulsory use of seat belts.

On June 11, 1981, the House of Lords met to debate the Transport Bill. Lord Nugent immediately moved to amend the bill to allow the secretary of state for transport to "make regulation requiring . . . persons who are driving or riding in motor vehicles on a road to wear seat belts."[30] The Lords voted 132 to 92 to accept Lord Nugent's amendment, and the bill was sent back to the Commons.

The debate in the House of Commons was scheduled on July 28, 1981, the eve of the royal wedding of Prince Charles to Lady Diana Spencer—a deliberate move, according to supporters of the amendment, to ensure poor attendance and endanger the seat-belt provision's chances. The timing of the debate may have hurt the pro-seat-belt side, but the political rule under which the bill was debated favored it. Since its first formal appearance in 1973, mandatory seat-belt legislation has been viewed as an issue of conscience, not subject to party discipline. Hence, both the Conservative and Labor parties allowed a "free vote" on the bill. Had the Conservatives, with their substantial majority, made the seat-belt bill a party issue, it would have suffered certain defeat, but following a lengthy debate, the House of Commons approved the seat belt amendment by a vote of 221 to 144. Of the 144 voting against the bill, 122 were Conservatives. However, 89 Conservatives voted in favor of the amendment and provided the necessary majority.

[29] *United Kingdom Parliamentary Debates* (Lords), vol. 421 (11 June 1981), col. 95.
[30] Ibid., col. 321.

According to observers, two factors contributed to the success of the measure. The first was the overwhelming and active support of various professional groups, including the police, the medical community, and academics. The second factor was that these groups were able to gain considerable attention from the media in their campaign to bring the issue before Parliament. Of particular influence was a widely viewed BBC documentary on automobile accidents entitled "The Greatest Epidemic of Our Time." According to one expert observer, "The program, viewed by almost half of the country because of its transmission twice at prime time, once the week before a crucial vote in Parliament, had a powerful effect on the result [of the vote]."[31]

Buckling Up in the United States

Not long ago a careful and highly respected student of public policy wrote that "in today's political climate mandatory belt laws seem unrealistic in the United States."[32] Seven years later, thirty-four states and the District of Columbia have adopted mandatory seat-belt laws. The reasons for this dramatic turn of events are not difficult to find. The first was passage in June 1984 of a New York State law requiring the use of seat belts by front seat occupants beginning January 1, 1985. In most areas of state policy innovation, once the legislative ice has been broken, subsequent policy debates and, often, adoptions follow quickly.[33]

The second and more important reason was the Department of Transportation's July 1984 decision to require either air bags or automatic seat belts (passive restraints) in all new passenger automobiles sold in the United States by 1990. The decision will be rescinded automatically, according to the ruling, "if states representing two-thirds of the U.S. population enact mandatory seat belt usage laws." By 1988 all but three states (Nevada, Idaho, and Kentucky) had taken some action on a mandatory seat-belt bill. However, although the trend clearly was toward increased adoptions, the issue was by no means resolved. In 1986 voters in Massa-

[31] Murray Mackay, "Legislation for Seat Belt Use in Britain," paper prepared for presentation at the Society of Automotive Engineers, International Congress and Exposition, Detroit, Michigan, February 27-March 2, 1984, 154.

[32] Kenneth E. Warner, "Bags, Buckles, and Belts: The Debate over Mandatory Passive Restraints in Automobiles," *Journal of Health Politics, Policy and Law* 8 (Spring 1983): 52.

[33] The seminal work in this area is Jack Walker, "The Diffusion of Innovations Among the American States," *American Political Science Review* 63 (September 1969): 880–89. See also Virginia Gray, "Innovations in the States: A Diffusion Study," *American Political Science Review* 68 (December 1973): 1174–85; and Robert Eyestone, "Confusion, Diffusion, and Innovation," *American Political Science Review* 71 (June 1977): 441–47.

chusetts and Nebraska approved ballot measures to rescind their states' mandatory laws, and in 1988 West Virginia Republican Governor Arch Moore vetoed a seat-belt law. In addition, most state belt laws do not meet the criteria established by the Department of Transportation to count in its calculation of the percentage of the population covered by mandatory legislation. These criteria include minimum fines of twenty-five dollars, exemptions only for medical reasons, adoption of education programs, and a provision that noncompliance may be used to reduce damages in civil lawsuits. Whatever their fate, seat-belt bills have generated considerable political discussion and controversy in the United States.

NEW YORK

In July 1984, New York became the first state to adopt a mandatory seat-belt law. Proposals for mandatory seat-belt legislation had been considered periodically by the state legislature for nearly a decade. Since 1980 New York had required the use of either car seats or seat belts by minors. Proponents of adult seat-belt legislation felt that the success of this measure provided a good opportunity to again propose a mandatory bill.

Interestingly enough, impetus for seat-belt legislation in New York came as a result of the defeat of a related measure. In May 1984 a bill that would have increased the state's drinking age from nineteen to twenty-one years was defeated in the Assembly. (In 1985, as a result of Federal prodding, the New York legislature approved the increase.) Defeat of the bill, which was seen primarily, although not exclusively, as a road safety measure, led a number of previously skeptical public officials, including Governor Mario Cuomo and Assembly Speaker Stanley Fink, to support a seat-belt bill. In addition, some state legislators who had opposed the raised drinking age favored the seat-belt bill because they felt it would not only save more lives, but also would not discriminate against any particular age group.[34]

The New York State Senate approved a seat-belt bill on June 13, 1984. The bill was amended in the State Assembly on June 21, where it was approved eighty-two to sixty, by just six votes more than were needed for passage, and it returned to the Senate for final approval by a vote of thirty-seven to twenty-two on June 25. Governor Cuomo signed the bill on July 12, 1984. It must be stressed here that the New York legislature approved the bill prior to Secretary Dole's decision on air bags. "Outside

[34] *New York Times*, 6, 20, and 22 June 1984.

pressure," which would play an important role in other states, was not a factor in the case of New York.

ILLINOIS

Illinois became the third state, following New York and New Jersey, to adopt a seat-belt law. Legislative opposition to the bill, however, was considerable, and it took over eight months for the measure to clear the Illinois General Assembly. The bill passed the state House of Representatives on May 16, 1984, by only one vote more than was required by law—sixty-one to forty-five. The bill was put on postponed consideration in the Illinois Senate until November. Thirty votes were necessary to pass the measure in the upper house, and on November 28 the bill was defeated by one vote, when one state senator was not in the chamber to verify his "yes" vote. A motion to reconsider was approved on December 12, and the Senate once again debated the seat-belt bill. This time, however, with the presence of three Democrats who were absent during the first vote, the bill was able to get two votes more than the minimum requirement and passed thirty-two to twenty-six.[35] Illinois Governor James Thompson, who did not take a position during the legislative vote, signed the bill on January 8, 1985. Illinois was one state in which "outside pressure" was felt in the legislative process. At the executive level, for example, Governor Thompson was lobbied by no less a luminary and automobile industry celebrity than Lee Iacocca. And, in the legislature the "hand" of the automobile manufacturers was, according to legislators, very much in evidence.[36]

OREGON

Bills to require seat-belt use in Oregon had been introduced as early as 1977. In that year, and again in 1981, the Oregon legislature rejected these efforts. It appeared, however, that in 1985 Oregon would join other early-adopting states. The 1983 legislature had approved a child restraint law, and the evidence that it was saving lives and reducing injuries was frequently mentioned by legislators and the press. Observers predicted easy passage of a seat-belt bill.

And, indeed, a seat-belt bill was quickly approved by the state Senate (nineteen to eleven) in January 1985. In early February the bill moved on to the House, where proponents predicted prompt approval. The bill immediately attracted twenty-five sponsors, just six short of the thirty-one

[35] See Illinois State Senate deliberations, 12 December 1984, 56 (hereafter IS).

[36] Ibid., 62; and *Chicago Tribune*, 29 November 1984.

votes needed for passage. However, in the one month it took to travel from the House Human Resources Committee to the full House, floor support for the bill eroded. By the time the House voted on March 14, supporters could only muster thirty votes. Eight of the original twenty-five sponsors had abandoned the bill. A last-minute effort to send the bill to the voters was rejected.

There is agreement among Oregon legislators and observers of the legislative scene that the bill was defeated because of an unorganized, grassroots effort that developed following Senate approval of the measure. Legislators began hearing from constituents in numbers large by Oregon legislative standards, and most of what they were hearing was negative; Oregonians did not like to be told by government that they had to use seat belts. Particularly vulnerable were first-term legislators in the House of Representatives. Five of the eight original sponsors who withdrew support were freshmen; of the remaining seventeen only three were first-term representatives. According to the bill's chief sponsor, Senator Rod Monroe, the first-termers simply were not "able to stand up to the pressure."[37] Monroe vowed to reintroduce a seat-belt bill in the next session of the Oregon legislature.

In terms of the seat-belt issue the 1987 legislative session was a carbon copy of the previous one. The Senate again handily passed a seat-belt bill (twenty to nine), and the House again narrowly defeated it (thirty-three to twenty-seven). The issue certainly would have died at this point but for one, ironical twist. Also pending before the legislature was a bill to increase the speed limit on rural freeways from fifty-five to sixty-five miles per hour, a measure many believed would result in increased highway fatalities. Governor Neil Goldschmidt and several legislators indicated that their support for the sixty-five mile-per-hour limit might be more forthcoming if a seat-belt bill were revived. A compromise of sorts was worked out. The legislature passed and the governor approved a mandatory seat-belt law for a one-year experimental period, submission of the issue to the voters in November 1988 along with a ballot measure to reimpose mandatory motorcycle helmet use, and an increase in the speed limit to sixty-five miles per hour. Opponents outside the legislature, however, were not yet reconciled to a seat-belt law, even on an experimental basis. In October 1987 a group opposed to mandatory seat-belt use collected over forty thousand signatures to prevent implementation of the law through a ballot initiative. Oregonians rejected mandatory legislation in November 1988, by a margin of 56 to 44 percent.

With this background in mind, it is now possible to compare the content of the debates over mandatory seat-belt legislation in each country.

[37] *Portland Oregonian*, 10 March 1985.

The Issues in the Seat-Belt Debate

Space does not permit a detailed analysis of all the factors that shaped the content and outcome of the seat-belt debates in these four cases. As already suggested, various idiosyncratic factors such as defeat of the raised drinking age in New York, the large number of freshman legislators in Oregon, and a television program in Great Britain, played some role in placing the issue on the legislative agenda or in shaping its fate. But none of these, in my judgment, was central to either the debates or their ultimate outcome. Two general concerns dominated. The first was a series of practical considerations, including questions of evidence; the second, and more important, involved fundamental political concerns, especially those involving the appropriate role of government.

EFFECTIVENESS OF BELTS

For the overwhelming majority of legislators, in both countries, there was simply no question that seat belts would save lives and reduce the likelihood of serious injury. Ivan Lawrence, Conservative M.P., and the most vociferous opponent of mandatory seat-belt legislation in Great Britain, said in the Commons debate, "We are not anti-seat belts. We are, on the contrary, overwhelmingly in favor of them. We agree . . . that the wearing of seat belts is far more likely than not to save the life of the driver or passenger in a car, and to avoid serious injuries."[38] In the same debate Norman Fowler, Secretary of State for Transport, an opponent of the bill, said, "I hope that nothing that is said tonight will challenge the proposition that seat belts provide protection in an accident and that it is a matter of common sense to wear them."[39] Nor did opponents of mandatory legislation in the United States question the efficacy of seat belts in saving lives and reducing serious injuries. According to an Oregon state senate opponent, "Certainly I think every one of us in this body today recognizes seat belts do in fact save lives. No one argues that case at all."[40]

Despite the overwhelming sentiment that seat belts save lives and reduce injuries, there were some who saw them as posing, under extreme circumstances, a threat to the motorist. Expressing a concern found on both sides of the Atlantic, an Oregon State senator said, "In a crisis situ-

[38] *United Kingdom Parliamentary Debates* (Commons) 6th ser., vol. 28 (28 July 1981), col. 1033.

[39] Ibid., col. 1048.

[40] Oregon State Senate deliberation, 7 February 1985, Senator Hannon (hereafter OS). Because the Oregon legislature records its debates on cassette tapes, not in printed form, it is not possible to list page numbers. Instead, the name of the legislator will be given for purposes of identification. Anyone interested in these tapes should contact the author.

ation they can become a death trap."[41] Those who expressed this concern referred to the dangers involved either in a fire or when an automobile became submerged in water.

Supporters of seat-belt legislation had little difficulty countering this argument. Evidence to the contrary was provided in a study of 20,000 automobile accidents in Wexham Park Hospital near Slough, England, which found that "in not one [case] was anyone damaged or inhibited by wearing a seat belt."[42] In addition, others pointed out how infrequently accidents involve either fire or submersion of a vehicle.

ENFORCEABILITY

The second "practical" issue involved the question of enforceability. "This law," according to one New York legislator, "is totally unenforceable."[43] An Illinois State senator protested that "it is impossible for a police officer to determine whether or not a motorist has on their seat belt," a sentiment expressed in virtually identical language by colleagues in Oregon, New York, and Great Britain.[44] Opponents asserted that as a result of the unenforceability of the law the life-saving and injury-reducing claims of proponents would not be fully realized because a significant proportion of the people would ignore the law.

British and American legislators were concerned with another facet of the enforceability issue, the implications the law would have for police-community relations. The connection was made by Ivan Lawrence, M.P.: "You had a situation where either the police weren't going to enforce it, or alternatively the police would try to enforce it and relations between the police and public would worsen."[45]

Although social scientists long have debated the reasons why people obey laws (for example, habit, fear of coercion, or normative value placed upon obedience), many lawmakers simply accepted obedience as a given. Proponents of seat-belt laws responded to this issue in identical fashion: our people are basically law-abiding. One New York assemblyman maintained, "Most New Yorkers and most Americans are law-abiding, and they are going to obey the law."[46] And David Ennals, Labor M.P., quoting

[41] OS, 7 February 1985, Senator Yee.

[42] *United Kingdom Parliamentary Debates* (Commons) (28 July 1981), col. 1063.

[43] New York Assembly (hereafter NYA), 13 June 1984, 4543.

[44] IS, 28 November 1984, 50; OS, 8 February 1985, Senator Meeker; New York State Senate, 13 June 1984, 4559, Senator Jenkins (hereafter NYS); *United Kingdom Parliamentary Debates* (Commons) (28 July 1981), col. 1050, Secretary of Transport Norman Fowler.

[45] Mr. Ivan Lawrence, M.P. (Conservative), interview by author. Tape recording, London, England, 4 June 1984.

[46] NYA, 21 June 1984, 143, Assemblyman Hevesi.

a letter to the London *Times* from the Association of Chief Police Officers of England, Wales, and Northern Ireland, said, "The law will to a great extent be self-enforcing as a large majority of the motoring public do not intentionally flout the law."[47]

"THE SLIPPERY SLOPE"

As important as these practical issues were, it was the political concerns that clearly dominated in each debate. For many opposed to mandatory seat-belt legislation, the essence of the issue was individual freedom. There were actually two, related, variations on the theme of the freedom-diminishing implications of mandatory seat-belt laws. The first of these was the "slippery slope." "Once you start this kind of thing there is no logical end to it and you are on a very slippery slope indeed."[48] This assertion was typically followed by such *reductiones ad absurdum* as "We can have the state police then arresting people for eating french fries or pastry or for not exercising enough,"[49] or "Every time we plug in our electric razor, I suppose we could get electrocuted. Are we going to outlaw electric razors?"[50] And, finally from one member of Parliament: "I have read articles by many experts that include advice that too much sex can cause heart trouble. . . . Let us have a law to say that we must not have too much sex."[51] The assumption is, of course, that erosions of individual responsibility and freedom are related and cumulative. Once the state replaces its wisdom for that of the individual, it is easier, and perhaps inexorable, that it will do so repeatedly.

An issue closely related to the slippery slope is the concern with delineating the area where the role the state plays in protecting public health and safety ends and the freedom of the individual to make life-style choices begins. This concern was articulated in virtually identical fashion by three legislators whose votes were cast thousands of miles apart geographically but who were philosophically adjacent. These legislators wanted to know: "Where do we stop?" "Where does government end and where should the individual take up just ordinary common sense?" "Where [do] we draw the line between the nanny state and the freedom of the individual to make sensible decisions?"[52]

[47] *United Kingdom Parliamentary Debates* (Commons) (28 July 1981), col. 1040.

[48] *United Kingdom Parliamentary Debates* (Lords) (11 June 1981), col. 354.

[49] Illinois House of Representatives (hereafter IHR), 16 May 1984, 216, Representative Johnson.

[50] NYS, 13 June 1984, 4548, Senator Cook.

[51] *United Kingdom Parliamentary Debates* (Commons) (28 July 1981), col. 1046, Mr. Lewis, M.P.

[52] NYA, 21 June 1984, 149; Oregon State House of Representatives, 13 March 1985,

The question of where to draw the line is both highly political and personal. What is viewed by one person as an appropriate function of government is viewed by another as an Orwellian intrusion. In general, proponents of mandatory seat-belt legislation tried to create the impression that it was neither an extraordinary departure from current road safety policy nor a significant intrusion on personal freedom. For example, British and American legislators sought first to establish that driving was already highly regulated. British M.P.s were reminded, "We know that there is no real freedom on the roads," and Oregon legislators that "the use of the automobile is the most regulated part of our lives already."[53]

Furthermore, proponents argued that there are literally hundreds of safety regulations, standards, and conditions governing motorists and their vehicles, each of which entails some loss of individual freedom for the sake of protecting the motorists and general public. Mandating seat-belt use too would require a loss of freedom, but this was characterized as a "minor infringement" (Mr. Jessel, British M.P.), "not a substantial intrusion" (Representative Currie, Illinois), and a "legitimate infringement" (Senator Bernstein, New York).

"SELF-REGARDING OR OTHER-REGARDING ACTION?"

The second facet of the individual freedom argument involves the distinction between what John Stuart Mill called "self-regarding" and "other-regarding" acts. According to Mill, "The sole end for which mankind are warranted, individually and collectively, in interfering with the liberty of any member of a civilized community, against his will, is to prevent harm to others. His own good, either physical or moral, is not a sufficient warrant."[54] Mill labeled those acts in which the individual brings harm to himself or herself, but not others, as "self-regarding" and therefore beyond the bounds of state interference. For some legislators in both Britain and the United States, not using a seat belt was such an activity. In the words of one M.P.: "Some ask what principle of freedom a seat-belt law would offend. My answer is the principle that we have a right not to be made into criminals if our actions are directly aimed at hurting only ourselves."[55] This sentiment was shared by American legislators as well: "Here we're talking about our own safety, our own lives, our own free-

---

Representative Day, (hereafter OHR) Mr. Peter Fry, M.P. (Conservative), interview by author.

[53] *United Kingdom Parliamentary Debates* (Commons) (28 July 1981), col. 1044; OS, 18 February 1985, Senator Monroe.

[54] John Stuart Mill, "On Liberty," in *The Philosophy of John Stuart Mill*, ed. Marshall Cohen (New York: Modern Library, 1961), 197.

[55] *United Kingdom Parliamentary Debates* (Commons) (28 July 1981), col. 1035.

dom and our own right to make our own conscious choices" (Representative Johnson, Illinois).[56] "When I'm going to do something that's going to hurt someone else, then enact your laws but when it's only me at stake, then forget about it" (Senator Schermerhorn, New York).[57]

Proponents of mandatory seat-belt legislation rejected the notion that the failure to use a seat belt produced no negative externalities. First, an unbelted driver is more likely to lose control of an automobile in an accident and involve other motorists or pedestrians. "The Bill would not only protect drivers and passengers in the front seat, the Bill would also protect other people. It would protect pedestrians on our highways and on our sidewalks . . . [when] a car is driven by a person who doesn't have a seat belt, even a minor injury can result in that person losing control of the car and injuring other people on [*sic*] or about the car" (Representative Cullerton, Illinois).[58]

Failure to use a seat belt imposes external costs in an even more direct sense; society must pay for this imprudent behavior not only in loss of lives but in loss of money as well. According to State Senator Roberts of Oregon, "We are saying that it is wrong to impose the penalties, the unnecessary penalties, upon individuals, their families, and society, not only because of the pain we suffer, but because of the economic effect we suffer."[59]

ISSUES UNIQUE TO AMERICAN DEBATES

The issues discussed thus far have been characteristic of seat-belt debates on both sides of the Atlantic. However, following the U.S. Department of Transportation's decision on air bags and passive restraints in July 1984, two new issues were introduced into the American debate. The first of these was economic; the second political.

After the Dole decision the automobile industry began a multimillion dollar national campaign—$12 to 15 million in 1985 alone—to promote state seat-belt legislation and thereby forestall the requirement for air bags or other passive restraints by 1990. The automobile manufacturers articulated their position during the hearings on motor-vehicle safety standards held by the NHSTA. During these hearings the manufacturers argued that consumers would object to automatic restraints and find ways to disable them, current manual lap and shoulder belts were more effective than passive restraints and air bags, and the additional costs of

[56] IHR, 16 May 1984, 217, Representative Johnson.

[57] NYS, 25 June 1984, 5236–37, Senator Schermerhorn.

[58] IHR, 16 May 1984, 212, Representative Cullerton; see also OHR, 13 March 1985, Representative Hooley.

[59] OS, 18 February 1985.

any passive restraint system, but especially air bags, would increase the costs of vehicles to the point at which industry sales, employment, and profitability would be adversely affected.[60] For these reasons, manufacturers decided to campaign for state seat-belt laws. Illinois legislators, for example, complained of being forced to "knuckle down" to the automobile manufacturers, while Governor Thompson, who met with Chrysler Corporation chairman Lee Iacocca shortly before signing the act, reported that Mr. Iacocca made "a very strong pitch" for the bill.[61]

Perhaps the greatest impact of the automobile industry, however, was in the Missouri seat-belt debate. The Missouri legislature approved a seat-belt bill in February 1985—and did so in great haste. The reason was that Missouri, along with several other states, was trying to convince General Motors to build its new, multi-billion-dollar Saturn automobile plant in that state. Governor John Ashcroft was scheduled to meet with General Motors officials on February 21 to make his pitch for the Saturn plant. State officials wanted the seat-belt bill passed prior to that meeting to demonstrate the state's pro-automobile-industry climate. In the course of the seat-belt debate it became clear that the issue of the plant overshadowed all others. According to Missouri House Speaker, Robert Griffin, "Right now the only thing coming to the top is Saturn. There's no talk about saving lives and preventing injuries."[62] According to Speaker Griffin, if the bill had not been linked to the plant, "it would not have any chance at all" in the House.[63] (Missouri, by the way, did not get the Saturn plant—Tennessee did.)

The second factor that has been present in American debates, but absent in the British case, was states' rights sentiment. Many state legislators condemned the fact that they were being forced into this decision by the federal government. Said one Illinois State senator, "It seems to me to be a melancholy fact that as we buckle up in Illinois, we must, at the same time, knuckle down to Washington . . . of all the reasons for passing a piece of legislation here in this body, I would say that pressure from Washington D.C. is one of the worst."[64]

It is difficult, based upon so few cases, to determine the impact of the "federal intrusion" issue on seat-belt debates. Mandatory seat-belt legislation has been debated in state legislatures for years, although never adopted. Certainly there is validity to Warner's assessment, noted earlier,

[60] See *Federal Register* 49: 28966, 28973, 28990. Automobile manufacturers have reluctantly come to accept air bags, which will become a standard feature on many cars beginning in the 1990 model year.

[61] IS, 12 December 1984, 59; *New York Times*, 28 February 1985.

[62] *Jefferson City Post Tribune* (Jefferson City, Missouri), 19 February 1985.

[63] Ibid., 14 February 1985.

[64] IS, 12 December 1984, 58.

about the general climate in the country being unfavorable to mandatory legislation. Clearly the Department of Transportation's decision forced the seat-belt issue onto more state policy agendas, and earlier, than would have otherwise been the case. Yet one should recall that the New York state legislature approved its bill prior to Secretary Dole's decision. My own feeling, based upon these cases and secondary sources from Missouri, Utah, North Dakota, and Washington, is that the issue of "federal intrusion" was, for most state legislators, only one part of a more general conservative response to government intrusion.

There is a remarkable resemblance between the British and American policy debates over road safety, especially safety belt and helmet issues. In both countries the dominant concerns, indeed the very language of the debate, was virtually identical. Contrary to other areas of health promotion policy discussed in this book, ideology and culture joined rather than separated the policy experiences of these two countries. In an intellectual and behavioral sense, the most fascinating part of the story was the ideological dilemma faced by conservatives on both sides of the Atlantic.

The British case is particularly relevant because of the insights it offers into many of the concerns raised by conservatives in the course of health promotion policy debates. To begin with, one condition for many of the eighty-nine British Conservatives who supported the seat-belt bill was the degree of certainty concerning the evidence that seat-belt use saved lives and reduced injuries. One Conservative told me, "There are areas where it is right for the state, with its knowledge and its power, to step in for the good of the community." However, this should only be done when there is "incontrovertible evidence that the behavior is harmful." This M.P. contrasted his support for mandatory seat-belt use, where to his mind there was such evidence, with his opposition to mandatory fluoridation of the water supply, where he thought the health evidence unconvincing.[65] Second, some Conservatives rejected the notion that freedom is indivisible and that the loss of one freedom undermines all freedom. As one member put it, "I don't believe in the thin edge of the wedge theory."[66] Finally, one Conservative M.P., Toby Jessel, argued that government intervention in this instance was not analogous to, say, an attempt to ban smoking or drinking. He believed that trying to prevent road deaths and injuries differed from trying to regulate other potentially self-harming ac-

[65] Mr. Roger Sims, M.P. (Conservative), interview by author. Tape recording, London, England, 18 June 1984.

[66] Sir Anthony Myer, M.P. (Conservative), interview by author. Tape recording, London, England, 4 June 1984.

tivities in that "if any of those things were banned or curtailed by law, the activity itself would be inhibited. However, the wearing of seat belts does not prevent anyone from driving where he wants, so it is not a parallel case."[67] The distinction made here, then, is between losing one's freedom to do something at all and losing one's freedom to do something in a particular way. For some, like Jessel, the latter case of modified freedom is an acceptable one.

Would such arguments be convincing to American conservatives? They would, I think, be far more unpalatable. The reason for this rests in the different conservative traditions in the United States and Great Britain. The most important of these differences is that British conservatism, with its historical ties to the aristocracy and monarchy, has a tradition of what Anthony Sampson calls "benign paternalism."[68] There has long been a tension within the British Conservative party between paternalism, born of noblesse oblige, and libertarianism, closely associated most recently with Margaret Thatcher. Enough Conservative backbenchers identified with this paternalistic strain to help pass the seat-belt bill. It should be remembered in this regard that the debate over seat belts came at the very beginning of the Thatcher years, when "wet" Conservatives still played a consequential role in the party. It is interesting to speculate whether the fate of the seat-belt bill might have been different in, say, 1985 or so. Would Thatcher have allowed a "free vote" at that time? Although the issue had been viewed as a matter of conscience since its first appearance in 1973, she has rarely been timid about departing from established practices.

American conservatism, unlike its British counterpart, has neither an aristocratic nor a paternalistic tradition. It is simply more difficult for American conservatives to accept state-sponsored health promotion legislation as something other than an unwarranted paternalistic intrusion on individual freedom. Americans, in this view, must be free to make their own life-style choices without the intrusion, benign or otherwise, of government.

Is there any justification then, in the minds of American conservatives, for the state to coerce people into behaving more responsibly? One answer is provided by C. Boyden Gray, who was counsel to the Bush Regulatory Relief Task Force under the Reagan administration. He argued, specifically with regard to seat belts, that failure to wear a seat belt "isn't a classic example of purely private behavior that has no external consequences. The whole question of highway accidents does put a very heavy

[67] *United Kingdom Parliamentary Debates* (Commons) (28 July 1981), col. 1061.

[68] Anthony Sampson, *The Changing Anatomy of Britain* (London: Coronet Books, 1983), 37.

burden on medical bills, which all taxpayers have to pay."[69] In economic terms, regulation is justified here because private action imposes costs on persons who are not parties to a transaction. For some conservatives, then, the externalities of allegedly irresponsible behavior may justify state infringements of personal freedom. In the United States, the debates over seat belts have revolved largely around the question of the nature and extent of the negative externalities; the issue of evidence has become largely irrelevant.

Similarly, since the 1960s the question of the etiology of harm has for the most part been absent from automobile safety policy debates. The general consensus in both countries is that more progress can be made in road safety by modifying driver behavior than in changing automobile or road design.

With these other issues largely resolved, there remains the question of how to accomplish what virtually everyone concedes needs to be done. Can greater seat-belt or helmet use, or a reduction in the use of alcohol before driving, best be achieved through education or regulation? Conservatives are, of course, predisposed toward inspiring greater prudence through education rather than requiring it through regulation. The evidence in the area of road safety suggests that education is not equal to the task. In Britain, the Conservative government was forced to acknowledge that education was not working to increase helmet use, and it opted for regulation. A significant number of Conservative backbenchers reached the same conclusion about seat belts. With the evidence concerning what constitutes dangerous road behavior so compelling, with the alternatives to regulation seemingly ineffective, and with increasing acceptance of the argument that road accidents impose substantial costs on the public, many British and American policymakers have reluctantly reconciled themselves to regulation.

What about the impact of structural factors on the direction and fate of road safety policy in the United States and Britain? The centralization of authority in Britain facilitated passage of this life-saving measure—it is estimated that the seat-belt law has saved about two hundred lives per year—while federalism has meant that citizens in sixteen U.S. states are as of 1990 not required by law to enhance their prospects of surviving an automobile accident. Depending upon one's ideological predisposition, this may or may not be a source of consternation. The point is that this case illustrates that the impact of formal political architecture on policy is not uniform. Health advocates in Britain have been hampered by political centralization in their efforts to deal with problems associated with

[69] Quoted in Kenneth B. Noble, "The Politics of Safety Has a Life of Its Own," *New York Times*, 15 July 1984.

smoking and alcohol abuse, but were helped in the case of seat belts, while just the reverse has been the case in the United States.

There is one final point to be made about British and American experiences in this policy area; political culture has differentially influenced policy consequences. The claim by legislators in both countries that their citizens are law-abiding and that the seat-belt law would be essentially self-enforcing has, not unexpectedly, proven more accurate in Britain than in the United States. I suggested in Chapter 1 that British political culture historically has been characterized as deferential; for the most part the British typically accept as legitimate even those laws that may be unpopular. As a result, seat-belt usage in Britain has consistently been around 95 percent. By way of contrast, Americans tend to pride themselves on their rugged individualism and a willingness to flout "unjust" laws. Although there is considerable variation among the states, and even within each state, in none is the user rate higher than about 70 percent—in 1988 Hawaii and Maryland led the nation with 71 percent and 73 percent respectively—and in many it is considerably lower—Utah had a user rate of only 29 percent and Nevada 31 percent. Political culture has had an important impact on the public acceptance of policy choices.

Seven

Dealing with AIDS: Just Desserts?

IN THE FALL of 1980 the United States Department of Health and Human Services published *Promoting Health/Preventing Disease: Objectives for the Nation*, which, as noted in Chapter 3, set national goals to reduce a wide variety of diseases by 1990. The objectives in the section on sexually transmitted diseases included reducing the incidence of gonorrhea, syphilis, herpes, and chlamydia; acquired immune deficiency or AIDS was not mentioned in the study. Within five years, AIDS was being described as "a major peril to our entire species," "one of the gravest public health threats of this century," and "the century's most virulent epidemic."[1]

By now the reader should have a sense that in the field of public health rhetorical restraint is virtually unknown. I have already quoted the most hyperbolic statements by public health officials and health promotion advocates concerning the consequences of smoking, alcohol abuse, and motorists' driving habits. Yet there *is* a nearly unique sense of urgency, in some instances bordering on hysteria, in professional and public discussions of AIDS. Rarely, if ever, in this century has any disease so captured popular fears, frustrations, and prejudices as AIDS. Rarely, indeed, has any disease emerged so rapidly and embedded itself so firmly in the national health policy landscape. Recent public health problems such as Legionnaires' Disease and Toxic-Shock Syndrome enjoyed similarly rapid infamy, but their impact was much more circumscribed in terms of numbers affected and duration of the etiological mystery surrounding them. AIDS is unlike any other life-style-related disease discussed thus far. The very categorization of AIDS as a life-style disease is itself controversial.

The impact of AIDS on British and American societies has been profound. In February 1987 the British Secretary of State for Health and Social Services invoked emergency powers last used during World War II when he directed government scientists to concentrate their efforts on AIDS research.[2] AIDS has reached into virtually every corner of private and corporate life: schools, the insurance industry, the arts, law enforce-

[1] See John Langone, "AIDS Update," *Discover* (September 1986): 30; Elizabeth Whelan, "The AIDS Epidemic," quoted in the *Congressional Record*, September 18, 1986, E 3150; and Robin Henig, "AIDS: A New Disease's Deadly Odyssey," *New York Times Magazine*, 6 February 1983.

[2] *Times* (London), 10 February 1987.

ment, housing, employment, and medical and sexual practices. It has influenced the way television and the movies portray life—even James Bond is no longer promiscuous—and the way we actually live it. In Britain, soccer players have been warned by their professional association not to kiss after scoring goals, and in the United States, "America's Team," the Dallas Cowboys, submitted to voluntary AIDS testing. In many American towns and cities police officers use rubber gloves when arresting gay protesters, and gloves have become as commonplace in dentists' offices as dental floss. No disease in this century, including the 1918–1919 influenza epidemic, and polio, has so transformed so many facets of our lives in such a short period of time as AIDS. And no disease has been so politicized.

Despite the infancy of the problem, clear differences have already emerged in the public policy responses in the two countries. These differences result from both social values and political culture; for example, in Britain there is relatively greater tolerance of homosexuality and relatively greater trust in and deference toward government. The consequence has been that public health policy dealing with AIDS has been better defined, coordinated, and seemingly successful in Britain than in the United States. Furthermore, political structure has influenced the policy response. In the United States the lack of direction and consensus from the national government, especially during the Reagan administration, hampered progress in dealing with the epidemic. Because microorganisms do not respect state boundaries, no amount of creativity and initiative at the state level can equal leadership from the federal government, which has yet to emerge in the United States. By way of contrast, the British government, supported by facilitating social and political values, was able to arrive at a consensus on how best to respond to the AIDS epidemic and move quickly to deal with it.[3]

The Nature of the Disease

In June and July 1981 the Centers for Disease Control (CDC) reported the results of a special task force looking into a surprising number of heretofore rare diseases. The June 5, 1981, CDC newsletter *Morbidity and Mortality Weekly Reports* noted five cases of a rare and generally fatal type of lung infection, pneumocystis carinii pneumonia, in young homosexual men in Los Angeles. The report raised the possibility of "an

[3] For quite a different conclusion concerning the policy responses to AIDS in Britain and the United States see Daniel M. Fox, Patricia Day, and Rudolf Klein, "The Power of Professionalism: Policies for AIDS in Britain, Sweden, and the United States," *Daedalus* 118 (Spring 1989): 93–112.

association between some aspect of homosexual life-style or disease acquired from sexual contact and pneumocystis pneumonia in this population."[4] The following month the CDC reported that Kaposi's sarcoma, a rare form of capillary cancer appearing just below the skin surface, which usually has been associated with elderly men of Mediterranean extraction, had been diagnosed in twenty-six young homosexual males over a two-and-one-half-year period. Both diseases had been found in individuals with severely depressed immune systems.

Over the course of the next several months, in scientific journals and meetings, additional cases of Kaposi's sarcoma, pneumocystis, and several other opportunistic infectious diseases were reported found among immune-suppressed homosexuals and intravenous drug users. The diseases appeared ultimately to be fatal in all instances. Early in 1982 the CDC labelled the new affliction "Acquired Immune Deficiency Syndrome." In technically unadorned terms, what appeared to be happening was that the immune systems of previously healthy individuals were breaking down, thus leaving the victims vulnerable to various infections or tumors, of which Kaposi's sarcoma and pneumocystis carinii were the most common. By early 1984 American and French researchers identified a virus, dubbed HTLV-III (Human T-Lymphotropic Virus Type III) by the Americans, and LAV (Lymphadenopathy Associated Virus) by the French, as the likely cause of the compromised immune systems. Ultimately American and French scientists agreed to call the virus HIV (Human Immunodeficiency Virus).

From this beginning, AIDS has grown in numbers and notoriety to become a worldwide health problem. The progression of the disease in Britain and the United States is noted in Figures 7-1 and 7-2, although public health officials in both countries candidly admit that the data on reported AIDS cases may be unreliable.[5]

The actual number of cases tells only a small part of the story. Perhaps the most obvious and chilling part of the disease's initial development was that the doubling time of the number of cases was between five and six months. The early trends seemed to portend a rapidly accelerating public health problem of alarming proportions. As recently as 1988 the United States Public Health Service was estimating that by 1992 over 260,000 people would die from AIDS, while in Britain the predictions ranged from nearly 7,500 to over 16,500.[6] More recently, however, there has been a

[4] United States Department of Health and Human Services, *Mortality and Morbidity Weekly Reports*, 5 June 1981.

[5] See the comments of Dr. Meade Morgan, AIDS Unit, Centers for Disease Control, in United Kingdom Department of Health and Social Security, *Future Trends in AIDS* (London: Her Majesty's Stationery Office, 1987), 21.

[6] Report of a Working Group, *Short-Term Prediction of HIV Infection and Aids in England and Wales* (London: Her Majesty's Stationery Office, 1988), 36.

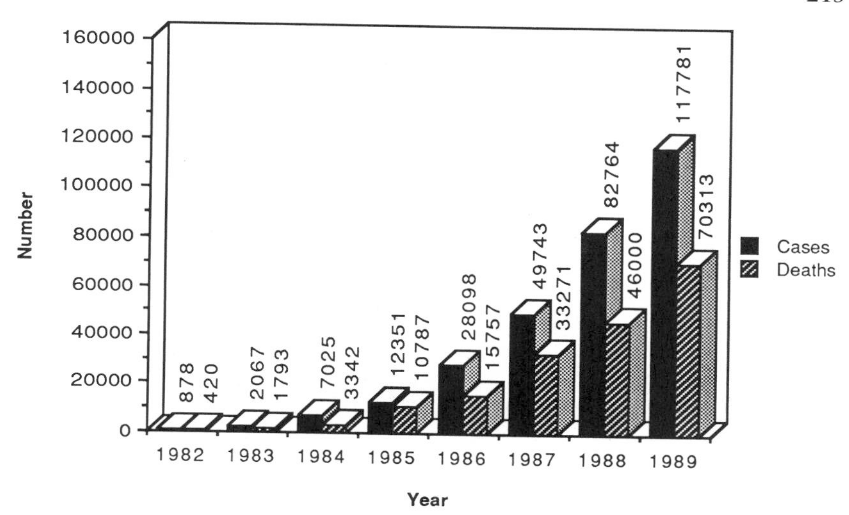

FIGURE 7-1
Cumulative Number of AIDS Cases, United States, 1982–1989
Sources: United States Bureau of the Census, *Statistical Abstract of the United States, 1989*, 110th ed. (Washington, D.C.: Government Printing Office, 1989), 113; *New York Times*, 11 February 1990.

sharp deceleration in the annual increase in the number of new AIDS cases. In 1989, for example, there was only a 9 percent increase in new cases over the previous year in the United States compared to 34 percent in 1988 and 60 percent in 1987.[7] Similarly in Britain estimates of new cases and deaths have been revised downward significantly. In 1988 a working group of the Department of Health and the Welsh Office predicted that in 1990 there would be 2,300 new cases and 1,370 deaths. A little over a year later the Public Health Laboratory Services estimated that by the end of 1990 there would be only 1,300 new cases and 750 deaths in Britain.[8]

Although many in both countries are now predicting that the worst is over in the AIDS crisis, no one was belittling the seriousness of the current and future problem. There are at least two reasons for caution about any predictions concerning the future trends of AIDS. The first is that no one knows with any degree of confidence how many people are infected with the AIDS virus. This, combined with the newness of the disease and the fact that there is an incubation period of between five and ten years, suggest that even short-term predictions must be treated gingerly. Second, in

[7] *New York Times*, 11 February 1990.

[8] Report of a Working Group, *Short-Term Prediction*, 36; and *Times* (London), 3 February 1990.

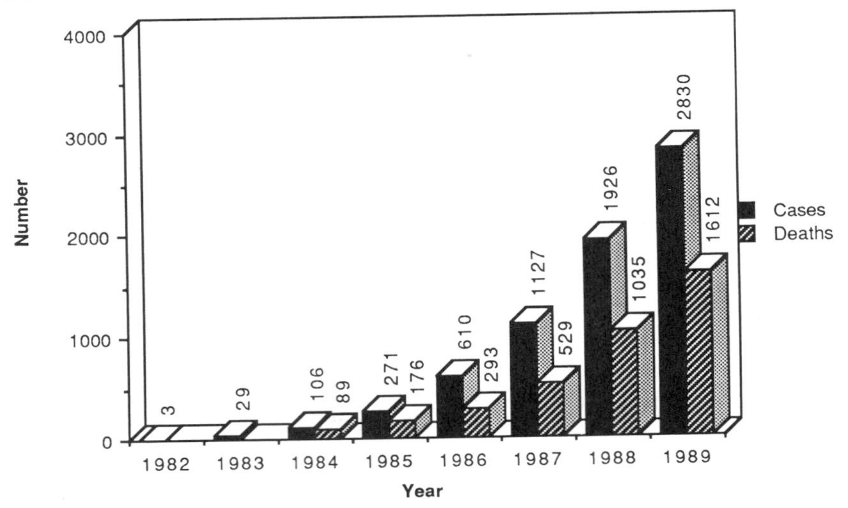

FIGURE 7-2
Cumulative Number of AIDS Cases, the United Kingdom, 1982–1989

Sources: London *Times*, 1 December 1988, 3 February 1990; Foreign and Commonwealth Office, *Survey of Current Affairs* 12 (London: Central Office of Information, 1988), 439.

both countries the slowdown in the spread of the disease has been greater within the relatively small male homosexual and bisexual populations than among heterosexuals. For example, the number of new AIDS cases among homosexual and bisexual males increased just 8 percent from 1988 to 1989, compared to a 27 percent increase among heterosexuals.[9] If the disease spreads in significant numbers to the heterosexual population, then the public health problem is certain to continue to be a grave one. It simply may be too early to determine the precise nature of the threat that AIDS poses to the public's health. One thing is clear: short of the discovery of a vaccine or successful treatment, AIDS will be a significant cause of illness and premature mortality in each country.

Because there are so many imponderables about the future course of the disease, no one is certain how much of a financial burden it will place on the U.S. and British health care systems. It is unclear how many AIDS victims there will be or what anti-AIDS drugs will be available. Currently only one antiviral drug, azidothymidine, or AZT, can prolong the lives of many AIDS victims and may delay onset of the disease in those who are infected with the virus but are asymptomatic. AZT costs about $8,000 a year per patient. (Another drug, pentamidine, has been found somewhat

[9] *New York Times*, 11 February 1990.

effective in staving off pneumocystis in those with already impaired immune systems.) Because of its high cost, and because many AIDS patients have lost private health insurance, about 40 percent of all U.S. AIDS victims eventually must turn to Medicaid for support. In Great Britain the entire expense must be borne by the National Health Service. Furthermore, experts are divided over whether the use of AZT will ultimately save money, by keeping victims out of the hospital longer, or add to the bill by prolonging their lives. In any event, estimates have placed the annual cost of treating the disease at about $5 billion, and in 1990 President Bush requested federal spending of $1.7 billion for AIDS care, research, and education. Projections for future direct costs range from $8.5 billion to $16 billion by 1991, and indirect costs are estimated as high as $55 billion for 1991. Even accepting the more modest figure for direct costs only will mean that by 1991 the national bill for the care of AIDS patients will exceed that of all other health claims except those of victims of automobile accidents. British authorities estimated that in 1988 AIDS cost the government more than £63 million ($81.3 million).

Despite the obviously foreboding message suggested by this data, there is little that is not controversial about the disease, including the nature and magnitude of the public health threat it poses.

A New Plague or The "Disease of the Week"?

The public health and moral implications of AIDS have prompted extraordinarily apocalyptic rhetoric. Cardinal Basil Hume, leader of Britain's Roman Catholics, described the spread of AIDS as a "moral Chernobyl," while the usually staid *Economist* announced in March 1987 that "the plague years have begun."[10] American observers have been no less dramatic in their characterizations. A secretary of Health and Human Services warned that an AIDS epidemic was approaching that would "dwarf" earlier epidemics such as the Black Plague, smallpox, and typhoid.[11]

The alarm expressed by scientists as well as political and religious lead-

[10] *Times* (London), 13 December 1986; *Economist*, 7 March 1987.

[11] The number of recent books on AIDS is expanding rapidly. See, for example, David Black, *The Plague Years* (New York: Simon and Schuster, 1986); Sandra Panem, *The AIDS Bureaucracy* (Cambridge: Harvard University Press, 1988); Susan Sontag, *AIDS and Its Metaphors* (New York: Farrar, Straus and Giroux, 1988); John Langone, *AIDS: The Facts* (Boston: Little, Brown and Company, 1988); Elisabeth Kubler-Ross, *AIDS: The Ultimate Challenge* (New York: Macmillan, 1987); Alan F. Fleming et al., eds., *The Global Impact of AIDS* (New York: Alan R. Iss, 1988); Mary Catherine Bateson and Richard Goldsby, *Thinking AIDS* (Reading, Massachusetts: Addison-Wesley, 1988); and Randy Shilts, *And the Band Played On* (New York: St. Martin's Press, 1987).

ers is apparently shared by the general public. A recent Gallup poll found that AIDS had replaced both heart disease and cancer as the most serious health problem in the minds of the American people; 68 percent identified AIDS, 14 percent cancer, and 7 percent heart disease as the nation's most serious health problem.[12] Similarly in Britain, a poll conducted for the London *Sunday Times* revealed that AIDS was viewed as the fifth most important problem facing the nation; only unemployment, nuclear war and disarmament, law and order, and the health service ranked higher.[13]

Two observations here about the public response to AIDS. First, reactions to public health problems in general tend to be rather mercurial, with the sense of concern highly correlated to the publication of government reports or a celebrity's illness or death. In light of the decline in the fear of widespread heterosexual transmission, and a deceleration in the increase in new AIDS cases, public anxiety, and the media attention that feeds it, also seems to be tapering off. One measure of this is that an October 1989 Gallup poll found that between 1987 and 1989 the percentage of those supporting legalization of homosexual relations increased from 33 to 47 percent. This latter figure represents a return to the 1982 pre-AIDS level of support for legalization.

Second, despite the intense and widespread concern over the public health implications of AIDS, not everyone agrees on the threat posed by this new disease. Dr. Albert Sabin, developer of the live oral polio vaccine, believes the "hysteria" surrounding AIDS is "unfounded." "The relative public health importance of AIDS is minuscule by comparison with many far more serious problems that make it look like a speck floating on the ocean. It has been blown out of all proportion."[14] The notion that the problem has been blown out of proportion is shared by a number of those who influence American public opinion. Nationally syndicated columnist Charles Krauthammer recently decried the money and attention devoted to AIDS, noting that "AIDS is not the pandemic its publicists would like us to believe, nor does it merit its privileged position at the head of every line of human misfortunes that make claims on our resources, attention and compassion. It is a disease. You would not know it from reading the papers, but there are others."[15] As noted earlier, because the disease is so new, and because there has not as yet been any systematic national sampling of a significantly large size population—a situation that will soon be rectified in Britain—epidemiologists and public health officials simply

[12] *New York Times*, 25 March 1987.

[13] *Times* (London), 1 February 1987.

[14] Quoted in John Langone, "AIDS Update: Still No Reason For Hysteria," *Discover* (September 1986): 32.

[15] *Portland Oregonian*, 15 June 1987; see also *New York Times*, 4 February 1987, and William Safire in *Portland Oregonian*, 7 June 1987.

do not know with any degree of certainty how prevalent AIDS is and what its future course might be. As one group of British specialists concluded, "Reliable data on the past and present prevalence of HIV infection would be the most effective starting point for predicting the incidence of AIDS. That route is largely barred by the paucity of seriological data."[16]

Differences of opinion over the public health danger posed by AIDS are more than simply varying interpretations of the epidemiological data, although that is part of the problem. Actually there are two issues involved here, one empirical, the other symbolic. The empirical issue involves an interpretation of the seriousness of AIDS compared to other diseases. This is in essence the point that both Sabin and Krauthammer have raised. For example, in 1987 Krauthammer reminded his readers that more people die each *month* from heart disease (65,000) than had died, in total, since 1981 from AIDS.[17] In Britain, 2,830 people had died from AIDS through 1989, while 100,000 die each year from smoking-related illnesses. No one is dismissing the very real tragedy of AIDS, but the question raised is whether the "hysteria" over the disease, and the concomitant demand for resources and attention are justified.

In particular, these critics ask, Is it justified given the epidemiological pattern of the disease? Thus far it would appear that AIDS remains largely within the known high-risk community; over 90 percent of the cases in both countries occur among homosexual or bisexual males, intravenous drug users, and a small and diminishing number of hemophiliacs. As long as there is no substantial outbreak into the heterosexual community, the chances of containing the disease are much brighter—and the political pressure for extraordinary action less pressing. By early 1990 the consensus among public health officials was that although heterosexual transmission of AIDS was on the rise, the disease would not spread in significant numbers to the non-drug-using heterosexual population. If this pattern holds, the expectation is that the number of AIDS cases within the gay community will level off at about 7,000 a year, but there will be an acceleration among intravenous drug users and especially those from the poor and minority communities. The changing demographics of the disease do not bode well for programs that depend upon health education and confidence in government institutions for their success.

In addition, lawmakers and public health officials face a dilemma in portraying the magnitude of the disease. On the one hand, the need to awaken the public to the dangers of AIDS, and to convince other public

[16] Report of a Working Group, *Short-Term Prediction.*

[17] *Portland Oregonian*, 15 June 1987.

officials to commit substantial resources to deal with the problem, would seem to require a fairly scary message. On the other hand, there is the danger of creating hysteria that might lead to the loss of the civil rights of AIDS victims and homosexuals. An aroused and frightened public might be willing to sacrifice the rights of a few in order to protect the health of the many. Congressman Henry Waxman (D-California), in a speech before the American Bar Association, sought to assuage some of the fears about the disease: "As hard as it is to imagine, AIDS is not as bad as it might have been. It is infectious, but it is not easily caught. This is not the bubonic plague of the Middle Ages. It is not the flu of seventy years ago. AIDS is not the worst case scenario for medicine or public health or good law."[18] The problem of the need to alert people to the dangers of AIDS without producing hysteria is faced by the British as well. A British Labor opposition spokesman in a House of Commons debate on AIDS noted that "in the past few months, as a nation we have gone from hardly talking about AIDS at all to scarcely talking about anything else. Without firm leadership there is a real risk of moving from complacency to hysteria as the fears and anxieties so often fed on ignorance gain momentum."[19]

The demographic scope of the epidemic has far-reaching political implications. First, there is the question of money. In an era of severe reductions in public research funding there is intense competition over relatively scarce resources. Those involved in AIDS research talk of the need for billions of dollars to fight the disease. This would mean, of course, new laboratories, sustained grants, and secure incomes for researchers over the next several years, something that those in cancer or alcoholism or arthritis research want and need just as badly. How the public and those who allocate public funds perceive the danger from AIDS will help influence these resource decisions. If it is true, or perceived to be true, either that the AIDS epidemic is leveling off or that it is largely confined to small and essentially pariah groups, there will be less compulsion to make extraordinary allocations to this research enterprise, or indeed to any other facet of the effort against the disease.

In the minds of many critics of both British and American policy, the conceptualization of AIDS as a "gay disease" or "gay plague" has had an important impact on official reaction to the disease. There is the lingering belief in both countries, but especially in the United States, that governments were slow to respond to the threat of AIDS because of the socially and politically marginal position of the primary victims. According to

[18] *Congressional Record* (13 August 1986), 2857.

[19] *United Kingdom Parliamentary Debates* (Commons), 6th ser., vol. 105 (21 November 1986), col. 808.

Stephen Jay Gould, "If AIDS had first been imported from Africa into a Park Avenue apartment, we would not have dithered as the exponential march began."[20] Similarly a British observer has said that AIDS "has tapped into society's natural homophobia. That has become an excuse for inertia."[21]

The relationship between the political and medical response to the disease and the nature of the population most affected is further complicated in the United States by the disproportionate incidence of AIDS among blacks and Hispanics. These two groups comprise 17 percent of the total adult population, but they account for 39 percent of all AIDS cases. In addition, over one-half of all women and three-quarters of all children with AIDS are black. Most observers attribute this higher incidence of the disease to the high intravenous drug use in those communities. Intravenous drug use is one of the major vehicles for transmission of the disease to heterosexual men and women.

Although some have claimed that the hesitant public response to the disease is characteristic of the poor treatment and attention paid to minorities in general in the United States, leaders of these communities acknowledge their own hesitancy in dealing with AIDS. John E. Jacob, president of the National Urban League, admitted that blacks feared a backlash if AIDS became too closely associated with race and that this "is one of the reasons the black community has been slow to address this issue, to put it on our agenda."[22] In addition, there have been moral and cultural obstacles to dealing more aggressively with the disease. According to Gilberto Gerald, head of minority affairs for the National AIDS Network, "Not only is homosexuality a taboo topic in both the black and Hispanic communities, but intravenous drug users have no political constituency either. AIDS patients are a minority within a minority."[23] Furthermore, among members of the Hispanic community there is the additional problem of the conflict between the public health message to use condoms and the Catholic Church's admonition against such use.

The extent to which AIDS is or can be characterized as a disease primarily of social outcasts remains of special concern to homosexuals. Their concern is fueled by the fact that some on the political and religious right in both countries have sought to portray AIDS as divine retribution for perverted sexual behavior. This theme dominated an editorial in the *Southern Medical Journal* by Dr. James Fletcher of the Medical College of Georgia:

[20] Stephen Jay Gould, "The Terrifying Normalcy of AIDS," *New York Times Magazine*, 19 April 1987.

[21] *Times* (London), 8 July 1986.

[22] *New York Times*, 2 August 1987.

[23] Ibid.

> [A] logical conclusion is that AIDS is a self-inflicted disorder for the majority of those who suffer from it. For again, without placing reproach upon certain Haitians or hemophiliacs, we see homosexual men reaping not only expected consequences of sexual promiscuity, suffering even as promiscuous heterosexuals the usual venereal diseases, but other unusual consequences as well. Perhaps, then, homosexuality is not 'alternative' behavior at all, but as the ancient wisdom of the Bible states, most certainly pathologic.[24]

Similar sentiments about the "wrath of God" and "the natural order of things" have been expressed in Britain. According to one member of the House of Lords, "I know of no other animal than mankind which indulges in homosexuality. I think it is fairly safe to assume that homosexuality, and very often promiscuity, are against the natural order of things, and therefore if persevered in may bring trouble, often on innocent and guilty alike. I do not think it is possible to get away from the wrath of God altogether, although you may not wish to call it that."[25] The notion that AIDS is a manifestation of God's disfavor with homosexuals is not restricted to the political fringes of American and British society. A recent Gallup poll found that 43 percent of those Americans surveyed believed that AIDS is "divine punishment for moral decline."[26]

The characterization of AIDS as a gay disease, in origin and communicability, threatens to reverse many of the gains homosexuals have won in recent years in terms of social acceptability, or at least tolerance, and legal protections. And there is evidence of both behavioral and attitudinal reversals. According to Gallup polls, opposition to legalization of homosexual relations between consenting adults in the United States rose from 39 percent in 1982, when AIDS first became publicly known, to 55 percent in 1987, although, as noted above, this trend seems to have been reversed, according to an October 1989 poll that found only 36 percent of respondents opposed to legalization.[27] Using a somewhat different measure of homophobia, the National Gay and Lesbian Task Force has offered as evidence the fact that reported incidents of anti-gay violence more than doubled between 1985 (2,042) and 1986 (4,946), nearly doubled again in 1987, when there were 7,008 reported anti-gay incidents, and increased again in 1988 to 7,200 acts of violence and harassment.[28] In Britain, too, homophobic sentiment became apparent in the first parliamentary debate on the issue, on November 21, 1986. Some members

[24] James Fletcher, "Homosexuality: Kick and Kickback," *Southern Medical Journal* 77 (February 1984): 150.

[25] Lady Saltoun of Abernathy quoted in *Times* (London), 14 December 1986.

[26] Reported in *Public Opinion* 8 (January–December 1986): 37.

[27] *Gallup Report No. 258*, March 1987, 13; and *New York Times*, 25 June and 28 October 1989.

[28] *New York Times*, 10 October 1987, 3 July 1988, and 8 June 1989.

traced the origins of the problem to the 1967 legalization of homosexual relations among consenting adults, and to those in Parliament "who actively promoted every form of [sexual] deviation."[29] Some Conservative members, in fact, introduced an early-day motion to repeal the 1967 Sexual Offences Act. This motion was largely a symbolic gesture, but it was just the sort of anti-gay activity with which homosexuals are concerned.

Biology or Social Pathology?

Controversy, or at least concern, surrounds not only the definition of the magnitude of the problem, but its nature and origins as well. There are two related concerns here: the link between voluntary behavior and a resulting harm, and the morality of the behavior. With regard to the first, on the face of it AIDS would appear to be a classic life-style disease. Over 90 percent of those with the disease are homosexuals, bisexuals, or intravenous drug users who engage in behavior that is known to involve high risk. With certain precautions the disease appears to be truly preventable. It is precisely because so many of those with AIDS contracted it through homosexual intercourse that some gays are concerned that AIDS not be defined exclusively as a life-style or self-inflicted disease. They would prefer a strictly medical definition and approach to the disease, and the moral impartiality implicit in such a definition. In no other health issue discussed in this book is the distinction between biology and social pathology so important. According to Dennis Altman, a political scientist who is gay, "Once questions of blame and responsibility for the disease intruded into public discussion, it was clear that AIDS would be political in a way that is unprecedented for a disease in modern times. The vehemence with which homosexuals were attacked for the disease . . . obscured the reality that this was a new and very dangerous epidemic disease for which no one could be held responsible in any real sense."[30] But held responsible they were. For Dr. James Fletcher, quoted above, and others, there is no doubt that AIDS is a life-style disease, etiologically analogous to diseases related to cigarette use and alcohol abuse. Actually, Dr. Fletcher has argued that just as physicians who champion the cause of disease prevention have sought to turn smokers into nonsmokers, and those who abuse alcohol into teetotalers, so too should they "seek reversal treatment for their homosexual patients."[31] In Britain, too, there is

[29] *United Kingdom Parliamentary Debates* (Commons), 6th ser., vol. 105 (21 November 1986), col. 834.

[30] Dennis Altman, *AIDS In the Mind of America* (Garden City, New York: Anchor Press, 1986), 25.

[31] Fletcher, "Homosexuality," 150.

widespread belief in the self-inflicted nature of AIDS. A recent national survey found that 57 percent of the respondents believed that "most people with AIDS have only themselves to blame."[32]

Interestingly enough, there are even some homosexuals who willingly subscribe to the life-style theory of AIDS. Writing in *New York Native*, a gay newspaper, Michael Callen and Richard Berkowitz argued against state efforts to change gay sexual behavior, and in favor of health education. In doing so they articulated a basically conservative position that should now sound familiar to the reader:

> We are not suggesting an end to promiscuity. *Ultimately, it may be more important to let people die in the pursuit of their own happiness than to limit personal freedom by regulating risk*. The tradition of allowing an individual the right to choose his own slow death through cigarettes, alcohol and other means is firmly established in this country; but there is also another American tradition represented by the Federal Trade Commission and the Food and Drug Administration which warns people clearly about the risks of certain products and behaviors.[33]

In this particular formulation of the "free to be foolish" position Callen and Berkowitz seem to ignore the issue of negative externalities, and particularly the question of the potential harm to others posed by homosexual promiscuity. The critical questions here, as in other life-style issues, are: How likely is it that your foolishness will bring harm to others? What is the nature of that harm? How credible is the evidence linking the behavior to the harm? I suspect that equating the particular kinds of promiscuous sexual behavior or drug use associated with AIDS with the use of cigarettes or alcohol is unacceptable to most people in Britain and the United States.

Characterizing AIDS as a purely, or predominantly, life-style disease will have a profound impact on public attitudes and ultimately the content of public policy. One of the hallmarks of the life-style movement is, of course, the assumption of individual choice and responsibility. To the extent that any health problem is attributed to personal irresponsibility, the public is more willing to accept the need to restrict personal choice. But AIDS is a public health and political problem of a different magnitude from other consequences of allegedly personal imprudence, in that the harm is not only self-inflicted but also the result of what are to many people morally repugnant acts. Altman and other homosexuals might dismiss as "nonsense" the notion that AIDS is the product of homosexual

[32] Lindsey Brook, Roger Jowell and Sharon Witherspoon, "Recent Trends in Social Attitudes," *Social Trends* 19 (London: Her Majesty's Stationery Office, 1989), 18.

[33] Michael Callen and Richard Berkowitz, "We Know Who We Are," in Altman, *AIDS*, 146. Emphasis added.

life-styles, but it is a view that the public, presented with fairly convincing and thus far consistent epidemiological data, does not dismiss so lightly.

There have been dissenting views, especially among members of the Conservative party, but for the most part the British studiously have avoided labeling AIDS a life-style disease. This deliberate policy decision, in part, may reflect a more tolerant attitude toward homosexuality in Britain than in the United States. Homosexual relations among consenting adults, in private, have been legal in Britain since 1967, whereas they remain illegal in nearly one-half of the U.S. states. In addition, the government apparently accepted the advice of public health experts that there was a greater likelihood of success for a health education campaign on AIDS if it were not cloaked in moral garb.

In fact, one of the striking features of the AIDS debate in Britain has been the strong influence that the medical profession has had in shaping public policy. This influence has led the Thatcher government away from a social pathological conception of the disease to a medical or biological one. The position of the medical community was articulated most forcefully during the 1987 hearings of the House of Commons Social Services Committee on problems associated with AIDS. In testimony before the committee, the BMA insisted on vigorous and explicit health education programs and messages. The BMA favored a campaign to encourage low-cost or free distribution of condoms to make them easily accessible. The association opposed universal mandatory testing and notification. With the exception of its own ambivalent attitude towards public support of free syringes and needle exchanges for intravenous drug users, the BMA's message was that AIDS policy should be morally neutral. On most major issues, including those noted here, the Social Services Committee and ultimately the government followed the recommendations of the BMA and other health professionals.[34]

The decision by the government to avoid moral judgments about sexual life-style choices, however, has not gone unchallenged. In December 1986 James Anderton, the chief constable for Greater Manchester and the new president of the Association of Chief Police Officers, spoke at the association's annual meeting. In the speech the chief constable labelled AIDS a "self-inflicted scourge of society" and those at risk as "swirling around a human cesspit of their own making."[35] The remarks set off a heated public discussion during which members of Parliament, local government officials, Conservative and Labor party members, and the press debated the propriety and perspective of Mr. Anderton's comments. At

[34] See House of Commons Social Services Committee, *Problems Associated with AIDS*, session 1986–1987 (London: Her Majesty's Stationery Office, 1987), especially 18 February 1987.

[35] See *Times* (London), 12 and 14 December 1986.

one point Thatcher spoke favorably of the chief constable's candor, if not the content of his comments.[36] Officially the government continued to avoid labeling AIDS a life-style issue and a homosexual disease—British billboards announce that "AIDS Doesn't Discriminate"—but many wondered if Thatcher's private feelings might not be somewhat different than her government's public policy. In addition, a group called the Conservative Family Campaign provided written testimony to the Social Services Committee in which they urged that AIDS victims be placed in isolation hospitals and that the 1967 Sexual Offences Act be repealed and homosexuality recriminalized. It is important to emphasize, however, that these views were decidedly in the minority.

In the United States, political leaders have been groping to find a politically acceptable course between the pressures from the right, which wishes to condemn homosexual life-styles, and from the left, which is concerned about protecting civil rights. In his first public speech devoted entirely to the subject—which he did not make until May 1987—President Reagan appeared to reject an attack on homosexual life-styles: "This is a battle against disease, not against our fellow Americans." Furthermore he called for "compassion, not blame," for AIDS victims.[37] The following day, Vice President George Bush echoed this theme in a speech before the largest scientific meeting yet held on AIDS. The vice president called for an all-out war against the disease, "not against the victims of AIDS." He went on to say that "we must remember that the sick and dying require our care and our compassion, no matter how the illness was contracted."[38] In a speech in March 1990 before the National Leadership Coalition on AIDS, President Bush reiterated this theme of compassion and attacked the victim-blaming mentality of some critics: "We don't spurn the accident victim who didn't wear a seat belt. We don't reject the cancer patient who didn't quit smoking cigarettes. We try to love them and care for them and comfort them. We do not fire them. We don't evict them. We don't cancel their insurance."[39] Nevertheless, some conservatives leave no doubt that how the "illness was contracted" is anything but irrelevant. George Will reminded his readers that: "America's principal public-health problems flow from *foolish behavior* regarding eating, drinking, smoking, driving—and, with AIDS, abuse of the body, especially the rectum."[40] (This is one instance, by the way, in which most conservatives do not feel people should be free to be foolish.)

The Reagan administration was torn between a moral and medical

[36] Ibid., 24 January 1987.
[37] *New York Times*, 1 June 1987.
[38] Ibid., 2 June 1987.
[39] *New York Times*, 30 March 1990.
[40] *Portland Oregonian*, 4 June 1987. Emphasis added.

conceptualization of AIDS. Its pro-family, anti-drug-use, conservative political instincts and fundamentalist religious bases of support predisposed it toward admonishing those segments of the population most at risk, and fashioning public policies that would discourage or penalize risky behavior. Yet the advice from the American medical establishment, including the president's own surgeon general, was to follow a less stigmatizing approach. This dilemma has faced policymakers in both countries during the short lifetime of the AIDS epidemic.

The Policy Response

Proposed policy responses to the AIDS epidemics in Britain and the United States have focused on two central questions: what is the best way to prevent the spread of the disease, and what is the most appropriate medical care and legal protection of those who already have the disease. I will be comparing the policy responses in the two countries throughout, but it is worth noting here that certain political and social differences between them have helped shape their overall responses.

The first, and most obvious, is that AIDS was discovered first in the United States, and it has involved a more substantial proportion of the American, than British, population: the 1989 rate of AIDS cases per 100,000 people in the United States was 48.9, and in Britain only 4.9. This has meant, in the first instance, that as in other policy areas, the British have tended to look to the U.S. experience in formulating their own policy responses. The November 1986 House of Commons debate and the 1987 Social Services Committee report were littered with references to the American experience, including suggestions, in the former, that the United States was the source of the problem itself. In addition, in January 1987 the Secretary of State for Social Services, Norman Fowler, visited the United States on a fact-finding mission. His trip included a stop at the Centers for Disease Control in Atlanta, a meeting with federal health officials in Washington, and visits to New York and San Francisco. A London *Times* headline announced, "Fowler visit to gay capital marks latest attack on Aids."[41] In an interview in New York, Fowler said, "What we're trying to do in touring the United States is to learn from your experience with AIDS and to see how we can apply it to our own experience."[42] The minister recognized that the problem thus far was less severe in Britain than in the United States and that one of the reasons for the trip was to keep it so. In addition, as mentioned in Chapter 1, when an advi-

[41] *Times* (London), 4 January 1987.

[42] *New York Times*, 25 January 1987.

sory group on AIDS prepared recommendations for the British Department of Health and Social Security on HIV-infected health care workers, it relied heavily on the U.S. Centers for Disease Control because of "their much greater experience of HIV in the health care setting."

Second, there is a significant difference between the two countries in terms of public attitudes towards homosexuals and homosexuality. Recent Gallup polls in both countries found that while 61 percent of the British supported legalization of homosexual relations between consenting adults, only 33 percent in the United States did.[43] Homosexual relations in Great Britain have been legal among consenting adults in private since the 1967 Sexual Offenses Act. That act was the culmination of a long, often bitter, debate that grew out of a report from a government-appointed Departmental Committee on Homosexual Offenses and Prostitution in August 1954. The committee, under the chairmanship of Sir John Wolfenden, submitted its report in 1957 and recommended decriminalization of homosexual relations.[44] It took ten years and parliamentary debates in 1960, 1962, 1965, 1966, and 1967 before the main recommendation of the Wolfenden Report became law in 1967.

It would be wrong to conclude that the formal legalization of homosexual relations, and continued strong public support for legalization, mean that homosexuals enjoy warm support in British society. A recent survey found that tolerance of homosexuals in education and positions of responsibility in public life increased between 1983 and 1987, but the study noted that such tolerance was not especially high, ranging from 40 to 50 percent.[45] Comments during the decade-long debate in Parliament—"In my opinion, in the general run the homosexual is a dirty-minded danger to the virile manhood of this country"—reflected continued revulsion toward homosexuality. Nevertheless, it is hard to avoid the conclusion that homosexuality has been a more visible, acknowledged, and openly discussed subject in Britain than in the United States. And, as public opinion surveys and legislative activity indicate, homosexuality does appear to be more widely accepted in Britain. This may be due, in part, to the oft-reported prevalence of homosexuality in such British elite public schools as Eton, Harrow, and Winchester, which must create some sympathy among the country's political elite for homosexuals. In addition, nineteenth- and twentieth-century British history provides a catalogue of famous personalities with achievements in the arts and public affairs who were known

[43] *Index to International Public Opinion*, 1984–85 (Westport, Connecticut: Greenwood Press, 1986), 561; *The Gallup Poll* 1986 (Wilmington, Delaware: Scholarly Resources, Inc., 1987), 215.

[44] See *The Wolfenden Report: Report of the Committee on Homosexual Offenses and Prostitution*, authorized American edition (New York: Stein and Day, 1963).

[45] See Brook, Jowell, and Witherspoon, "Recent Trends," 17.

or thought to be homosexuals or bisexuals. This list includes Oscar Wilde (who was convicted in 1895 of "gross indecency"), T. E. Lawrence, Cecil Rhodes, Field Marshall Lord Kitchener, Somerset Maugham, and John Maynard Keynes.[46]

There is no direct, comparative evidence that gays in Britain are more readily accepted into society than their American counterparts, but the legalization of homosexual relations and the prominent role of homosexuality in elite education do seem to suggest and explain a generally more tolerant social attitude. In a series of four surveys conducted by the government between February 1986 and February 1987, between 72 and 80 percent of the respondents expressed the view that "homosexual men should be judged on their own personal merits just like everyone else," although around one-half said that they would "feel embarrassed if I learnt one of my friends or relatives was homosexual."[47] One of the important political consequences of this more socially tolerant attitude is that gays probably have greater access to policymakers in Britain than in the United States. As Fox, Day, and Klein note, "The gay lobby was less visible in Britain than in the United States, perhaps because it had fairly easy access to the [health] department."[48] In addition, the legality of homosexual relations means that certain policies are more practical in Britain. For example, one of the reasons why gays in the United States have opposed mandatory reporting of AIDS cases is that to do so in the twenty-five U.S. states that still have anti-sodomy laws would be tantamount to admitting to engaging in an illegal activity.

Third, policy issues and options in the two countries differ in part because of the difference in the health delivery and social insurance systems. Among the issues created by the AIDS epidemic in the United States has been eligibility for health and disability insurance, both public and private. Because of the universal nationalized health and social insurance system in Britain, there simply is no question about whether or not AIDS patients are eligible to receive health and disability assistance.

Despite these differences, there is one important political thread joining the two countries. AIDS descended upon British and American societies at a time when both were governed by leaders with strong personal commitments to traditional values, including emphasis on family life, sexual chastity, and abhorrence of drug use. These values, although somewhat stronger in the United States, with its history of moral puritanism and the

[46] For a thorough discussion of homosexuality in modern British history, see H. Montgomery Hyde, *The Love That Dare Not Speak Its Name* (Boston: Little, Brown and Company, 1970).

[47] *AIDS: Monitoring Response to the Public Education Campaign, February 1986–February 1987* (London: Her Majesty's Stationery Office, 1987), 82.

[48] Fox, Day, and Klein, "The Power of Professionalism," 98.

stronger influence of religious fundamentalism, often made it difficult for leaders in both countries to accept, with any degree of enthusiasm, policies that appeared to tolerate "deviant" social practices. It would appear that Thatcher's own personal discomfort in this area has been held in check by more tolerant British social attitudes towards homosexuality, the major influence of the medical establishment, and the comparatively less severe nature of the AIDS epidemic in her country. Thus far, at least, British policy has been both more consistent and enjoyed more consensus than American policy. In the United States, President Reagan's own ambivalence, and divisions within his administration, resulted in substantial policy indecision and conflict, a problem that the Bush administration has yet to resolve.

Prevention

PUBLIC HEALTH EDUCATION

There is little disagreement, in either country, that until someone finds a cure or vaccine, the best way to control the spread of AIDS is to educate the public to the dangers, causes, myths, and means of prevention of the disease. Echoing the words of the U.S. surgeon general, Norman Fowler declared that "public education and information, is basically the only vaccine we have."[49] Beyond this point, however, agreement breaks down, at least in the United States. Foremost among the issues in the political debates over the health education campaigns have been the nature, content, and audience of the health education message.

The single most contentious of these issues, especially in the United States, has been content. There are basically two competing perspectives on what the message should be. One counsels sexual abstinence outside of marriage and fidelity within. Those who advocate this also favor an educational approach that is uncompromising in its position on drug use: the only acceptable message is "don't do it." The second perspective advocates "safe sex," which has been characterized as or understood to mean avoidance of multiple sexual partners and sodomy. However, if neither of these admonitions is followed, then the use of condoms is the best way to minimize the danger of spreading the disease. In the case of intravenous drug users, the message is not to share drug-use equipment.

The differences between these two perspectives are important in terms of both political philosophy and public policy. The first reflects a fairly uncompromising moral position about sexual practices and drug use. The message, in effect, is that certain things are immoral, they can be fatal,

[49] *Times* (London), 4 January 1987.

and the best way to escape this consequence is to avoid the offending behavior. The second perspective is a pragmatic one. Its proponents reconcile themselves to certain social practices without intentionally endorsing them, although critics argue that this is precisely what they are doing. It urges policies, proponents argue, that will work rather than simply make a moral statement: moral compromises such as these are necessary because a message promoting safety is more likely to be effective than one urging abstinence. This is a position taken by the Institute of Medicine/National Academy of Sciences (IOM/NAS), which advocates "factual educational programs designed to foster behavioral change." The IOM/NAS recognizes that "this may mean supporting AIDS education efforts that contain explicit, practical, and perhaps graphic advice targeted at specific audiences about safer sexual practices and how to avoid the dangers of shared needles and syringes."[50]

The British government, apparently with some reluctance on the part of Thatcher, has opted in favor of the pragmatic safe sex and safe drug-use message. Although a number of members of Parliament, Chief Constable Anderton, and the Catholic Church have opposed sanctioning immoral behavior, the overall safe sex strategy has not caused the degree of conflict that it has in the United States. In November 1986 Thatcher appointed a special Cabinet committee to advise the government on AIDS policy. The committee included some of the most powerful people in the government, including the home, social services, foreign and defence secretaries, and it was chaired by Lord Whitelaw, the Lord President of the Council and Leader of the House of Lords. The government provided £20 million ($29.4 million) in 1987 for a "safe sex" health education campaign. The campaign was to be highly explicit—Thatcher is reported to have said that the campaign message was more suitable for lavatory walls than government publicity—and was intended to reach the entire nation.[51] One of the main features of the campaign was a leaflet entitled "Aids: Don't Die of Ignorance," which was mailed in January 1987 to each of the 23 million households in the nation. The leaflet warned that because the message deals "with matters of health and sex you may find some of the information disturbing. But please make sure that everyone who may need this advice reads this leaflet." It then went on to give some fairly explicit advice—for example, semen taken in the mouth might be risky, water-based lubricants for condoms were preferred over oil-based

[50] IOM/NAS, *Confronting AIDS: Update 1988* (Washington, D.C.: National Academy Press, 1988), 651.

[51] John Warden, "Letter from Westminister," *British Medical Journal* no. 293 (29 November 1986); see also, *Times* (London), 17 October 1986.

ones, and if you do inject drugs intravenously, don't share equipment.[52] In addition to the leaflet, the effort included newspaper and television advertisements ("Your Next Sexual Partner Could Be That Very Special Person: The One Who Gives You AIDS") a national poster campaign ("AIDS Is Not Prejudiced: It Can Kill Anyone"), and a magazine, radio, and movie campaign aimed at young people. Here too unprecedented explicitness was in evidence; condoms have been shown pulled on bananas, the tongue of an actor, and a plastic penis.[53] Finally, the government took the position that AIDS education had to reach into the primary schools and sent copies of a factual booklet "AIDS: Some Questions and Answers" to all primary as well as secondary school teachers.

If those in government have been forced to compromise on moral standards with regard to condoms, those standards are tested even more severely on the issue of whether or not the government should provide free needles and other paraphernalia to drug users. This issue has posed a genuine moral and political dilemma for the Thatcher government, which has waged a vigorous anti-drug-use campaign. Although the BMA was noncommittal on the issue during its testimony before the Social Services Committee, other public health and education experts have advised the government that providing free needles and syringes would be the most effective way of reducing the spread of the disease, especially to the heterosexual population. Already uncomfortable with the "safe sex" approach, and its implicit acknowledgement that people will engage in "promiscuous" sex, the government now had to decide whether it wanted to appear to promote, and even facilitate, "safe drug abuse." In fact, some local government-run hospitals had begun trade-in programs for drug users, and in early 1987 the government announced that it would set up fifteen needle and syringe exchange centers around the country.

The Social Services Committee recommendation reflected the continued national ambivalence over this issue. "We recommend that the needle and syringe exchange scheme be extended to other areas of the UK only if and when there is sufficient evidence that such schemes can be effective." This conditional support was endorsed by the government in its response to the committee recommendations.[54]

Needle exchange programs have had little success or support in the

[52] United Kingdom Department of Health and Social Security, "AIDS: Don't Die of Ignorance" (London: Her Majesty's Stationery Office, 1986).

[53] For a review of the impact of this public health education effort, see *AIDS: Monitoring Response*.

[54] House of Commons Social Services Committee, *Problems Associated with AIDS*, xxxiii, and; United Kingdom Department of Health and Social Security, *Problems Associated With AIDS: Response by the Government to the Third Report from the Social Services Committee*, session 1986–87 (London: Her Majesty's Stationery Office, 1988), 13.

United States. Even the IOM/NAS gave qualified support for the "evaluation of the effectiveness of providing sterile needles and injection equipment to drug abusers in certain circumstances."[55] At the beginning of 1989 only Tacoma, Washington, had an ongoing needle exchange program, while New York City had a very limited project. In other cities, including Los Angeles, Chicago, and Boston, such proposals have been opposed by members of the minority community who feel it simply would encourage drug abuse. The Bush administration, following the path of its predecessor in being unable to speak with a single voice on AIDS-related issues, was split on exchange programs. Bush's Secretary of Health and Human Services, Dr. Louis Sullivan, supported the idea but would leave it up to each community to decide if it wanted such a program. William Bennett, Bush's "drug czar," opposed exchange programs and any federal support for them. The president, of course, had the final word on the issue, and he "is opposed to the exchange of needles under any condition."[56]

The question of condoning or tolerating gay sexual practices and intravenous drug use haunted the Reagan administration in its struggle to develop a coherent and effective public health education campaign. It was not until May 1988, nearly one and a half years after the British mail campaign, that the surgeon general mailed brochures to each American household warning of the dangers of AIDS. Although everyone agreed that it was necessary to educate the public, there was considerable disagreement over the content of the educational message, particularly as it was addressed to young people. The AIDS pamphlet ("Understanding AIDS") was not nearly as explicit as its British counterpart: it contained only a brief section on condoms.

Within the administration the two alternative health messages, safety versus chastity each had a forceful proponent. On one side of the debate was Surgeon General C. Everett Koop, a conservative on abortion and other family-related health issues but the administration's main proponent of explicit instruction on safe sexual practices, especially within the schools. In particular, the surgeon general advocated explicit instruction on the use of condoms as the most effective barrier, short of abstinence, against the spread of the disease.

The other position was articulated by Reagan's Secretary of Education, William J. Bennett, and had the vocal support of conservative activists. Bennett championed the cause of sexual restraint among young people and responsible sex within marriage as the cornerstones of the general health education message. Furthermore, in contrast to Dr. Koop, who

[55] IOM/NAS, *Confronting AIDS*, 86

[56] See *New York Times* 9 and 11 March 1989.

believed sex education should begin as early as kindergarten, Secretary Bennett believed in postponing classroom discussions of sex until junior high school, and then only in the context of marriage.

In the spring of 1987 the Reagan administration's policy position began to emerge. The outline of the policy was contained in a memorandum to the Domestic Policy Council from Attorney General Edwin Meese. The highlights of the policy were (1) "there should be an aggressive Federal effort in AIDS education"; (2) the federal government should provide local governments and schools information on AIDS but the actual content of the school curriculum should be "locally determined" and "consistent with parental values"; (3) government health information "should encourage responsible sexual behavior based on fidelity, commitment and maturity, placing sexuality within the context of marriage," and it "should teach that children should not engage in sex."[57] The president acknowledged in April that his views were close to those expressed by Bennett. While Dr. Koop continued to urge that safe sexual practices be taught in the schools, it was left to local school districts and state and local governments to decide the content of sex education and AIDS prevention instruction.

Health education efforts are not, of course, restricted to schools. State and local governments have developed programs to reach high-risk communities, especially gays, prostitutes, and intravenous drug users, in the United States the message and commitment varies. As one might expect, areas with especially high rates of AIDS cases and homosexual populations have responded most imaginatively to the problem. Yet it is also in cities like San Francisco and New York, with large, well-organized, and politically active gay communities, that the dilemma of balancing health and moral considerations has been most troubling for the homosexual community. In San Francisco and New York, some of the anxiety of the gay community has been reduced through cooperative health education campaigns by local health departments and gay organizations. In contrast to Britain, where the central government has set a national standard for the health education campaign (although with the support of such private groups as the Terrence Higgins Trust), one can expect continued debate and acrimony in the United States over the appropriate health education message, as states and communities fashion a message that will satisfy both public health and local definitions of public morality.

DISEASE CONTROL AND CONFIDENTIALITY

Both countries have placed a high priority on controlling the spread of AIDS through health education. Although there has been controversy,

[57] *New York Times*, 26 February 1987.

most notably in the United States, over the content of the health education message, there is universal agreement on the value of this approach. No such agreement in principle exists in other areas of proposed AIDS containment policies. Lurking behind most of the other disease prevention options is the now-familiar conflict between community health needs and individual privacy and the protection of personal rights.

Foremost among these policy options is the identification and tracking of the AIDS-infected population. From a public health perspective there is a critical need to monitor the spread of any infectious disease. Indeed, modern epidemiology is predicated on the collection and availability of just such data. In the case of AIDS, data is needed on two populations: those with confirmed cases of AIDS whose immune systems are compromised, and those who harbor the virus, as measured by a test for the AIDS antibody, but who are asymptomatic. It is not known precisely how many of those who test positive for the HIV antibody will eventually get the disease. Initially it was thought that between 20 and 30 percent of those who tested positive developed the disease within five years, and 50 percent within nine years. The current conventional wisdom, and fear, is that virtually all of those who test positive may eventually develop the disease.

Data on asymptomatic virus carriers is important to predict the magnitude of the disease, provide an opportunity to prevent or treat various opportunistic infections associated with AIDS—something that new treatments are making increasingly possible—and to control the spread of the disease. If it is ultimately decided that to contain the disease it is necessary to register and monitor those with the virus, this information will be vital. The issue is, of course, quite sensitive for this very reason. For both those with AIDS and those who test positive for HIV, the question of who should know about their condition and what should be done with this information raises anew the conflict between individual rights and the protection of public health. To the extent that citizens trust public officials to respect their right to confidentiality, health surveillance procedures are more likely to enjoy support. The British political culture is more likely to inspire such confidence than the American value system.

Since the earliest days of the AIDS epidemic, the CDC and its British counterpart, the Communicable Disease Surveillance Centre, have required that confirmed cases of AIDS be reported to health authorities. In the United States, this means that hospitals, laboratories, physicians, and other health professionals must report cases of AIDS to local public health authorities. Today all states require reporting confirmed AIDS cases, although specific reporting requirements are set by each state. Notification is a standard practice for infectious or communicable diseases, but even this issue became controversial with regard to AIDS. Even though more than one-half of the states now require reporting of the names of those with confirmed cases, there remains considerable variation in reporting the HIV antibody test. New York, California, and Massachusetts, for example, prohibit or severely limit test disclosure, while about one dozen states require it. Members of the gay community are particularly concerned that disclosure of the names of persons with AIDS or the HIV antibody will lead to discriminatory practices in employment, housing, education (especially of children with the disease), insurance, medical care, and even mortuary services.

Through legislation, judicial decisions, and administrative action, about half of the states and the federal government offer some protection from civil rights violations of AIDS victims, but there is enormous variation among the states. More than one dozen states and the federal government bar discrimination in employment for persons with AIDS, typically under statutes that prohibit discrimination against those with mental or physical handicaps.[58] In the case of the federal government, this protection was modified by the Reagan Justice Department. In June 1986 the Department of Justice ruled that AIDS victims were protected from job discrimination under the 1973 Rehabilitation Act only if it could be proven that bias resulted from the victim's handicapped status. If, however, employers refused to hire or wanted to fire a person to prevent the spread of the disease, it would not be considered discriminatory. Furthermore, the Department of Justice ruled that simply being seropositive for the AIDS antibody was not a condition covered by civil rights legislation and, therefore, could presumably be the basis for rejection for a job or housing. This subtle, but obviously important, distinction reinforced the view of many in the gay community that they were vulnerable and that maintaining confidentiality was critical to the protection of their rights. This concern was strengthened when the Reagan administration opposed a proposed federal law, sponsored in 1987 by Congressman Henry Waxman (D-California), which would have guaranteed confidentiality to persons who had been tested or received counseling for AIDS. The secretary of health and human services argued that civil rights guarantees were most appropriately the concern of state and local governments, not the federal government. A year later the president's own AIDS commission—officially called the Presidential Commission on the Human Immunodeficiency Virus Epidemic—also endorsed strong new federal laws to prevent discrimination and guarantee confidentiality to protect those who test positive for the AIDS virus.[59] Rather than issue an executive order or endorse antidiscrimination legislation, President Reagan ordered federal

[58] For a review of some of the state actions in this area see Gene W. Matthews and Verla S. Neslund, "The Initial Impact of AIDS on Public Health Law in the United States—1986," *JAMA* 275 (16 January 1987): 347–350.

[59] IOM/NAS, *Confronting AIDS*, 64.

agencies to adopt guidelines to prevent discrimination against HIV-positive individuals in the workplace. This was viewed by many as the weakest of the three possible ways of dealing with discrimination: executive order, legislation, or guidelines.

With the development in 1985 of a test to detect the AIDS antibody, a broader, potentially more effective and controversial approach to tracking and containing the disease presented itself. It was now possible to identify those infected with the AIDS virus who had not, and might never, develop a full blown case of AIDS or AIDS-related complex. This new test, which virtually eliminated one source of the spread of the disease, namely, infected blood in the nation's donor blood banks, introduced additional issues that are still being debated. In addition to the persisting question of confidentiality, there was now the issue of whom to test and for what purposes and uses the test results would be applied.

There is virtual unanimity that wider testing is necessary to better track and predict the course of the disease, as well as to counsel those infected regarding prevention of spreading the virus. By 1990, federal prisoners, would-be immigrants, blood donors, military recruits and personnel, and certain other federal employees had to be tested. In addition, there have been various proposals to test people treated in hospitals, drug abuse and venereal disease clinics, and penal institutions. Within the Reagan administration, William Bennett advocated a more aggressive position, calling for mandatory testing of high-risk groups such as prostitutes and drug users, as well as marriage license applicants. Bennett received support from conservative Congressman William Dannemeyer (R-California) who since 1985 has sponsored legislation that would require testing for the groups recommended by the education secretary, as well as requiring the states to report on those persons who test positive for the AIDS antibody.

Dannemeyer joined forces in 1988 with Paul Gann, a leader of the California 1970s tax revolt, to win approval of a state referendum (Proposition 102) that would allow insurance companies and employers to require a test for AIDS, as well as require doctors to both report to health officials the names and addresses of persons who test positive for HIV and notify sexual partners. Gann had contracted AIDS as a result of a transfusion during heart surgery in 1982. (He died in September 1989.) The measure, which had the support of California Republican Governor Deukmejian but was opposed by public health and education officials and the gay community, was defeated in the November 1988 general election by a two-to-one majority.

Opposition to proposals for widespread mandatory testing involves several practical, public health, and philosophical objections. One practical objection is the ever-present issue of cost. Although there are no precise figures on the costs of a large-scale national testing program, the price for such tests ranges from twelve to three hundred dollars each, with an average of about seventy dollars. When applied to specific cases, the costs become impressive. For example, in July 1987, when the Illinois legislature approved a bill requiring the testing of prison inmates and hospital patients, it was estimated that it would cost $60 million to test prison inmates and between $9 and 12 million annually to pay for testing of patients who could not afford private testing. One state legislator called the proposal "nutty."[60]

Opponents of mass testing programs have argued that the millions of dollars spent on identifying a small number of cases would be better spent on increasing voluntary test sites, education, and counseling. As long as most AIDS cases involve those in high-risk categories, testing of low-risk groups is a very costly way of identifying a relatively small number of cases.

One interesting example of the problems that mass testing faces occurred in the state of Illinois. In 1987 the state adopted a law that went into effect on January 1, 1988, and required marriage license applicants to show proof of having been tested for AIDS. It was not necessary, however, to provide the test results. The law proved a boon for neighboring states: in an effort either to avoid the $30-$125-per-person test, or to make a statement about the invasion of privacy, Illinois couples flocked to neighboring states to be married; the number of Illinois marriage licenses fell by 25 percent from 1987 to 1988, and the number of Illinois couples seeking marriage licenses in Kenosha, Wisconsin, increased from fifty in 1987 to fifteen hundred in 1988. More importantly, during the eighteen months it was in effect the testing program revealed just forty-four positive HIV results out of a total 221,000 tests. The cost to Illinois couples was $5.4 million, or about $209,000 for each detected case.[61] In June 1989, at the urging of state health officials and AIDS specialists who had originally opposed the prenuptial testing, the law was repealed. (Louisiana, which had experimented with a similar law, repealed it after six months.)

In general, public health officials tend to endorse the objections of civil libertarians to selective and mandatory testing policies. In addition to the cost-benefit objections they are concerned that mandatory testing will drive underground many of those most vulnerable to infection, including intravenous drug users and homosexuals (especially in those twenty-five states with anti-sodomy laws), thus undermining the utility of the tests.

There is, in addition, another practical medical objection to large-scale testing. Although the test has an accuracy rate of 99 percent, even a 1

[60] *New York Times*, 21 September 1987.

[61] Ibid., 25 June 1989.

percent error when applied to millions of people would produce some serious problems. There is the danger in large-scale testing of producing a significant number of either "false positive" results, which would unnecessarily terrify some people, or "false negative" results, which would mean some people would unknowingly feel free to continue sexual behavior that posed a continuing danger to their sexual partners.

Opponents of mandatory testing have proposed instead widespread but voluntary testing, with effective and credible assurances of confidentiality. In a May 1987 report, the CDC recommended a program that would give increased access to, and encouragement of, testing for such high-risk groups as people being treated for sexually transmitted diseases, those with a history of intravenous drug use, pregnant women who may be at risk because of their sexual history, homosexual and bisexual males, prostitutes, and hemophiliacs. Voluntary testing, in combination with counseling programs, is the best way, according to the CDC, of informing those most at risk and persuading them to modify their behavior. Furthermore, the CDC urged the states to adopt or strengthen laws protecting confidentiality and penalizing discrimination of persons with AIDS or those who have tested positively for the antibody. This approach recommends itself from both an epidemiological and civil libertarian perspective. The CDC urged that counseling programs emphasize the need of those who have tested positively to inform their sexual partners. This should be done on a voluntary basis, but the CDC recognized that state and local health authorities may have to inform sexual partners of infected persons without permission.

Perhaps more than any other, the issue of confidentiality reveals the tension between protecting the community's health and the individual's civil rights. Many have argued that individuals and businesses have the right and need to know that the people who handle their food, teach their children, purchase their insurance policies, or with whom they have any sort of close or intimate contact are free of a disease or virus that is typically fatal. Despite the medical evidence that HIV can not be transmitted through casual contact, the public fear of the disease is such that appeals to rationality have been inadequate. In addition, information concerning the incidence of the disease, including more precise data on the actual number of cases and sexual partners involved, and its demographic and geographic distribution, is essential to containing the spread of AIDS. But, according to Dr. James O. Mason, former director of CDC, "Right now we are paralyzed. We don't have the data to fight this epidemic simply because we don't dare test large populations for fear of violating confidentiality and being charged with discrimination and invasion of privacy."[62]

[62] Ibid, 10 February 1987.

On the other hand, both the abstract principles involving personal rights and the specific realities of the legal and social vulnerability of gays, undocumented aliens, and drug users are equally compelling reasons to guarantee privacy and confidentiality. At a minimum, confidentiality is necessary to insure the acceptance and, therefore, effectiveness of screening programs. In one federally financed study of AIDS by Columbia University's School of Public Health, 410 of the 745 homosexual males asked to participate refused to do so. According to the research team's principal investigator, Dr. John Martin, the main reason given was fear of disclosure: "No matter what we told the people, they believed that if insurance companies or the Government wanted to get their hands on their names, they could."[63] As one member of the Gay Men's Health Crisis in New York put it, "In theory, tracing may make sense, but in practice it does not work. The fear and loathing and discrimination surrounding AIDS would drive people underground."[64] It is very unlikely, and understandably so, that a homosexual will submit to screening if there is any prospect that nonanonymous positive results will be shared with employers, insurance companies, or even government agencies, thereby risking their jobs, education, housing, or personal security.

The Presidential AIDS Commission recommended a federal statute to guarantee confidentiality of test results in order to prevent discrimination: "An effective guarantee of confidentiality is the major bulwark against that fear."[65] Neither President Reagan nor Congress supported such a law. In October 1988, when Congress passed perhaps the most significant AIDS legislation to date, involving a three-year expenditure of $1.5 billion for education, research and anonymous testing, as well as the creation of a national AIDS commission, it deleted a provision, sponsored by Congressman Henry Waxman and supported by the public health community, which would have assured confidentiality of AIDS tests.

We have, of course, seen the tension between limiting individual freedom and promoting public health in other areas of health promotion policies. But AIDS is not quite like other life-style-related diseases. It deals with far more private and intimate personal behavior. The implications of this are underscored by Ronald Bayer:

> But, in contrast to the difficulties that would be posed by efforts to limit the transmission of HTLV-III infection, those presented by attempts to modify smoking, alcohol consumption, and vehicular behavior are simple. In each of these cases we could, if we chose to, affect behavior through product design, through pricing and taxation mechanisms, through the regulation and control

[63] Ibid., 1 March 1987.

[64] Ibid., 27 January 1987.

[65] Ibid., 3 June 1988.

of essentially public acts. Invasions of privacy would be largely unnecessary. With the transmission of HTLV-III the public dimension of the acts that are critical for public health is exceedingly limited.[66]

It is clear that in the future a larger proportion of the American population will be tested for presence of the AIDS antibody. Currently there is considerable variation among and even within states over AIDS testing. An increasing number of states and localities are adopting measures that while trying to insure privacy will allow medical professionals to inform the family members and sexual contacts of antibody-positive individuals. The testing policy options available to decision makers will be influenced, in part, by medical developments. Should the disease break out into the heterosexual population, there will be greater demand for mandatory, universal, reportable testing. Similarly, if improved long-term treatment or a cure becomes available, wider testing will make better medical sense, and enjoy greater acceptance, than is currently the case.

The British debate over screening for the AIDS antibody and confidentiality of test results has not generated nearly the amount or intensity of attention that it has in the United States. The Whitelaw Committee accepted the advice of the British medical community that widespread testing was unnecessary and counterproductive, and that confidentiality was essential to voluntary screening and counseling efforts. Medical confidentiality was also endorsed by the Social Services Committee and by the government in its response to the committee's report. There is, in fact, precedent for such an approach in Britain. During World War I, government fear of a syphilis and gonorrhea epidemic led the British to adopt a program of free, walk-in clinics in which anonymity was assured. This practice has continued for other venereal diseases and was updated in the National Health Service (Venereal Disease) Regulation of 1974. In effect, the British have applied these regulations to AIDS. As described by the chief medical officer of the Department of Health and Social Security:

These regulations require that, when an infection has been sexually transmitted, employees of a health authority may not disclose information to persons not involved in the treatment of the patient without the patient's consent. In seropositive cases the doctor will advise the patient how to avoid transmitting the infection to others and also recommend that the sexual partner or partners should be told. The doctor has discretion to inform a third party to prevent the spread of infection.[67]

[66] Ronald Bayer, "AIDS, Power and Reason," *The Milbank Quarterly* 64, supplement 1 (1986): 172.

[67] E. D. Acheson, "AIDS: A Challenge for the Public Health," *The Lancet* (22 March 1986): 665.

Once the government accepted the medical community's judgment that confidentiality was necessary to wage an effective public health education and behavior modification campaign, it was easier to win the confidence of even homosexuals and drug users than it will be in the United States. This is not to deny that concerns over confidentiality, as well as discrimination in employment, housing, insurance, and other areas do not exist among the British gay population, but merely to make the comparative observation that the level of civic trust in Britain is almost certainly higher than in the United States.[68]

In November 1988 the government announced two new screening programs to monitor the spread of HIV in the population. The most important was anonymous screening in which blood taken from hospital and clinic patients for tests other than for AIDS would be analyzed for the HIV virus. Patients would not be informed of the results since the samples would not be identified. The program of anonymous testing was accepted by the Government after the BMA, the Royal College of Physicians, the Health Education Authority, and the AIDS Policy Unit all advocated it as the best way of accurately and reliably gauging the prevalence of the virus and the future course of the disease. The second program involved named screening of pregnant women in Dundee and Edinburgh, Scotland—Scotland has a higher incidence of drug-abuse-related AIDS cases than other parts of the United Kingdom. The secretary of state for health said the named screening program would ultimately be extended to England. Those tested would be volunteers and would be informed of the results if they wished.[69]

Finally, in January 1990 the Department of Health began the most ambitious screening scheme to date when it introduced a nationwide program of anonymous testing in several antenatal and sexually transmitted disease clinics, and general hospitals. Blood samples will be drawn for other diagnostic purposes but will also be tested for HIV antibodies. Patients are informed that they may choose not to have the AIDS analysis, that they will not be told the results, and that anonymity is guaranteed. The program, which is to be permanent, is intended to provide the most accurate assessment available on the prevalence of HIV infection among the general population. The health department expected that up to 250,000 people will have been screened by the end of 1990.

[68] For statements of concern about breaches of confidentiality, see the 25 February 1987 "Minutes of Evidence Taken Before" House of Commons, Social Services Committee, Session 1986–1987, 112–13.

[69] See United Kingdom Department of Health and Social Security, *AIDS and Drug Misuse Part I: Report by the Advisory Council on the Misuse of Drugs* (London: Her Majesty's Stationery Office, 1988), 55–59.

QUARANTINE AND OTHER MEASURES

It is not surprising, given the nature and seriousness of the disease, that AIDS has prompted proposals for far more aggressive public health measures than either education or screening. And it is perhaps not surprising that just as the disease itself has prompted many to call for a return to more "traditional" values of fidelity and, presumably, heterosexuality, so too has it encouraged demands for more "traditional," and restrictive, public health responses. Some have proposed, for example, the use of quarantine to control the spread of the disease. (The reader will remember that the British Contagious Diseases Acts empowered authorities to register prostitutes, require them to take a medical examination, and detain them for a specified period of time.)

Current interest in quarantining stems from the same fear and frustration in dealing with earlier contagious diseases. Proposals have ranged from the most outrageous—a Nebraska psychologist who wanted to quarantine all homosexuals—to more modest measures allowing health authorities to forcibly detain AIDS patients who are unwilling to discontinue behavior that places others at risk. Under the 1984 Public Health (Control of Diseases) Act on disease control, the British government gave local magistrates the authority to forcibly hospitalize persons with AIDS who appeared to be a risk to others.[70] The first, and only, use of this authority resulted in public disclosure of the victim's name and aroused considerable condemnation. The provision has not been enforced since. Similar action has been taken in parts of the United States. In a recent case in Florida a court ordered a prostitute with AIDS confined to her home, while Los Angeles County allows health officials to use involuntary detention against those with AIDS who might knowingly infect others. Finally, Illinois allows quarantine when there is strong evidence that a person willfully and knowingly is transmitting a sexually transmitted disease and poses a public health threat.

In both countries, however, there have been demands for broader quarantine powers. A right-wing British pressure group called Conservative Family Campaign has urged that the government isolate all AIDS victims, and in California over 680,000 people signed a petition to place a proposition on the ballot in November 1986 that would have declared AIDS an "infectious, contagious, and easily communicable disease."[71] The California measure was proposed by an organization called PANIC (the Prevent AIDS Now Initiative Committee), sponsored by right-wing political

[70] Michael Mills, Constance Wofsky, and John Mills, "Special Report: The Acquired Immunodeficiency Syndrome," *The New England Journal of Medicine* 314 (3 April 1986): 934; *New York Times*, 7 August 1987.

[71] *Times* (London), 9 January 1987.

activist Lyndon LaRouche. The proposition, which was defeated, would have allowed, and in some circumstances required, public health officials to test anyone suspected of having AIDS, to prohibit them from attending or teaching schools, and to quarantine those with the disease as well as those with the virus antibodies.[72] Although the measure was opposed by most reputable public health officials, scientists, and physicians in the state, and was defeated by a two-to-one margin, it was symptomatic of the fear that the disease has generated.

There are a number of reasons why the use of quarantine to control the spread of AIDS is both medically and politically questionable. First, there are the practical considerations of the cost associated with testing, maintaining, and supervising up to 1.4 million carriers of the AIDS virus in the United States. Second, quarantine historically has been used for diseases that have had finite infectious or communicable periods, the one exception being leprosy. Carriers of the AIDS virus, however, are infectious for their entire lives. It is one thing to quarantine someone for one month; it is quite another to do so for years. It is also well to remember that unlike transmission of any of the previous infectious diseases, the spread of AIDS requires some fairly extraordinary activities, thus making quarantine a less justified approach to containment.[73]

Beyond these practical considerations lie the considerable political difficulties that widespread use of quarantine would create. Even allowing for the comparatively weak political power of homosexuals, it is difficult to imagine either British or American policymakers adopting measures, at least for significant numbers, which would result in the loss of some of our most fundamental civil rights. Although there have been compromises before between the public health needs of the many and the political rights of the few, quarantine of large numbers of AIDS victims, or antibody-positive individuals, would constitute a compromise of democratic values of an unprecedented nature—perhaps the internment of Japanese-Americans during World War II would be the closest analogy.

The political difficulties posed by highly restrictive public health measures was illustrated in the United States by the brief but bitter controversy over another highly restrictive AIDS measure, the effort by the San Francisco department of health to permanently close that city's homosexual bathhouses. The bathhouses, which have been described as " 'erotic ghettos' where men go for high-volume, anonymous sex," seemed an ideal venue for the rapid transmission of AIDS among San Francisco's gay population.[74] A debate over the health and political implications of clo-

[72] Charles Petit, "California to Vote on AIDS Proposition," *Science* 234 (October 1986): 277–78.

[73] Bayer, "AIDS, Power, and Reason," 176.

[74] *Wall Street Journal*, 8 February 1985.

sure raged not only between the gay community and health officials, but among gays themselves. Many homosexuals opposed closing the bathhouses, fearing that it was a portent of further assaults on gay life-style, activities, and business enterprises—in other words, a variant on the "slippery slope" theme. One protester at a San Francisco press conference carried a sign reading, "Today the tubs, tomorrow your bedrooms."[75] Others argued that it was not the location of the high-risk sexual behavior but the behavior itself that was the critical issue. Closing the bathhouses would not change behavior; only education would do that, and the bathhouse was a better place than most to educate gays to the dangers of certain sexual practices. And, of course, there were civil liberty and individual choice arguments against closure. Sounding a theme that in a different context might have come from Orrin Hatch or Strom Thurmond, the president of the American Association of Physicians for Human Rights wrote, "The closing of businesses to protect people from themselves cannot be accepted. We must try to educate people there as well as elsewhere—and indeed it may be easier in bathhouses—but ultimately each individual is responsible for himself."[76] Once again, people should be free, even to be foolish.

Convinced, however, that the educational message aimed at the gay community in general and bathhouse patrons in particular was not having the necessary impact, in October 1984 the director of the San Francisco public health department, Dr. Mervyn Silverman, ordered the closure of fourteen bathhouses that were operating under conditions that fostered the spread of AIDS. Ignoring the order, all reopened virtually immediately, prompting the city to seek an injunction against the owners. Several weeks later the California Superior Court in San Francisco ruled that the bathhouses could remain open but must operate under conditions that would inhibit high-risk sexual activities—for example, partitions were removed so that all bathhouse activity would be open to view and monitoring, and patrons engaging in risky sexual behavior would be required to leave.[77]

After he had left the San Francisco health department, Dr. Silverman coauthored an article for the *Hastings Center Report* in which he articulated some of the public health and political questions raised by the bathhouse controversy:

[75] Altman, *AIDS*, 152.

[76] Quoted in ibid., 152.

[77] For discussions of the bathhouse controversy see ibid., 147–55; Mervyn Silverman and Deborah Silverman, "AIDS and the Threat to Public Health," *Hastings Center Report* (August 1985): 21–22; Mills, Wofsky, and Mills, "Special Report," 935; and Ronald Bayer, "AIDS and the Gay Community: Between the Specter and the Promise of Medicine," *Social Research* 52 (Autumn 1985), 595–97.

> Does the government have the right, for health purposes, to close an establishment whose structure is not inherently unsafe or unsanitary? To close, is it sufficient only to establish an association between behavior that occurs therein and disease? What about the paradoxical situation in which the government of San Francisco is spending millions of dollars to reduce spread and treat victims while commercial establishments are profiting from the promotion and facilitation of dangerous behavior? When does health regulation supersede the rights of consenting adults? What about the rights of the business-owners? What constitutes safe sexual behavior and how should it be regulated?[78]

Clearly some of these questions are limited to the issue of bathhouse closures, but others touch upon the more fundamental concerns dealing with the conflicts among public health, political rights, life-style choice, and disease prevention or control. The tension, for example, between the economic rights of bathhouse owners and public health is similar to that in the tobacco, alcohol, and even automobile industries. Certainly the question of when public health regulations should supersede the decisions of consenting adults pertains to all areas of life-style-related legislation. Yet one must come back to an earlier point, the particular social and political configuration of AIDS does make it different from other issues. The fact that it involves our *most* intimate life-style choice, groups whose lifestyles are unacceptable to a majority of people, and a fatal disease in which the link between behavior and harm is irrefutable, makes AIDS unprecedented in terms of the degree to which it has become politicized.

A compelling case can be made for the uniqueness of the AIDS epidemic. In fact, Sandra Panem, in what is perhaps the best political analysis of the problem to date, *The AIDS Bureaucracy*, categorizes AIDS as a "novel health emergency." She ably and convincingly describes why the U.S. public health establishment was so unprepared when confronted with the epidemic. A summary of Panem's analysis is worth quoting at length because it so nicely captures the myriad of the public and private intra- and inter-organizational obstacles to a responsible, coordinated, and effective response to AIDS.

> The timely resolution of a novel health emergency requires closely coordinated efforts by diverse institutions. Yet given the way the system functions, normal everyday operations make an efficient and integrated response to a new disease extremely difficult. Business as usual—includes adversarial relationships between executive and congressional branches of the federal government, rivalries

[78] Silverman and Silverman, "AIDS and the Threat," 22.

among different agencies within the Department of Health and Human Services, lack of clear-cut lines of authority and accountability among officials in different levels of government, and informal relations between the public and private sector—is part of the problem.[79]

Panem calls for setting up special procedures and plans to "liberate" future managers of novel health emergencies from the bureaucratic morass that undermines efforts to effectively respond to public health crises. Although I wholeheartedly endorse Panem's conclusions, I think they tell only part, albeit an important one, of the U.S. policy response to AIDS. A comparative perspective on U.S. AIDS policy reveals additional problems that may not lend themselves so readily to the institutional engineering suggested by Panem.

First, as I have suggested throughout this chapter, although homosexuals may not enjoy a privileged place in British society, in terms of legal protection, historical and cultural role, and elite and popular attitudes, theirs appears to be a somewhat more secure and accepted role than that of their American counterparts. Added to this was a health profession that ably and promptly laid out a judicious and responsible plan for dealing with the educational, therapeutic, and research dimensions of the AIDS crisis, and with which the government largely concurred. Indeed, as Daniel Fox, Patricia Day, and Rudolf Klein demonstrate, the British government was intent upon "creating and maintaining political consensus" on an approach to the problem. "This consensus rested on the repudiation of the view that AIDS could be seen as a 'gay plague,' a belief that the solution lay in changing the entire population's sexual behavior through public education and a rejection of measures such as compulsory testing, which might threaten civil liberties as well as be ineffective."[80] This, in turn, has facilitated a better-coordinated and more effective response to the AIDS epidemic. The British have begun widescale anonymous testing, have a significant drug paraphernalia exchange program, and have a national health education program that is far more candid, and one suspects effective, than the long delayed and more sexually circumspect American effort. Widespread homophobia in the United States would make it difficult to find a consensus on resource allocations, legal rights, or any other facet of any gay-related problem, and not just a novel health emergency.

Second, and related to this, because the British political culture continues to be characterized, to a considerable extent, by deference to and trust in political authority, policies that are viewed with nearly pathological suspicion in the United States are seen as less threatening, even by ho-

[79] Panem, *The AIDS Bureaucracy*, 136.

[80] Fox, Day, and Klein, "The Power of Professionalism," 96–97.

mosexuals, in Britain. As citizens in the state of Illinois demonstrated in the case of prenuptial AIDS testing (and as citizens across the country did with regard to mandatory seat-belt laws), Americans often see public health efforts as a challenge to beat the system rather than as a concerned attempt to protect their well-being. Suspicion of governmental interference in our personal lives hampers the policy response to AIDS and other public health problems.

My own cross-national and cross-policy approach to understanding AIDS policy diverges from Panem's because of another, implicit, reason. AIDS would have provoked an extraordinary popular and public policy response regardless of when it emerged. The fact that it did so during the time of American, and to a lesser extent British, interest in other life-style and health issues helped color popular perceptions and political reaction to the epidemic. In the context of the "new perspective" on health, AIDS can be viewed either as the ultimate affirmation of the deleterious consequences of personal irresponsibility, or the ultimate consequence of the victim-blaming bias inherent in that perspective. As such, the AIDS epidemic provides almost a caricature of the consequences and implications of the new perspective on health policy. Especially telling is the tendency to politicize health policy. There is, of course, a political dimension to virtually every health issue involving the allocation of public funds and the regulation of individual and corporate behavior in the name of public health and safety. But in the case of AIDS the political dimension, especially in the United States, has overwhelmed most other considerations. The ultimate consequences of defining AIDS as much politically as biologically have been profound. According to Dennis Altman, "There has been resistance on the part of most politicians and journalists to see AIDS as a public health crisis rather than as the disease of promiscuous homosexuals, who somehow infect innocent victims with the illness."[81] Society has the right to limit personal freedom, in the name of the common good, when individuals voluntarily engage in activities that threaten the health and safety of others. AIDS is perceived primarily as the result of an irresponsible life-style choice rather than, say, a random biological occurrence. Although the emphasis on personal choice suffuses the new perspective on health, nowhere is the message of victim-blaming more pronounced than in the case of AIDS.

Not only is the illness-provoking behavior considered imprudent, and other-regarding, but it is, to many, morally repugnant. The concept of irresponsibility, so central to the new perspective, sometimes merely implies a morally neutral recklessness. Although there has been a certain

[81] Dennis Altman, "The Politics of AIDS," in *AIDS: Public Policy Dimensions*, ed. John Griggs (New York: United Hospital Fund, 1987), 24.

degree of opprobrium attached to tobacco and alcohol use or abuse, AIDS is associated with behavior that is unacceptable to many, probably most, in British and American societies. The accusation that the new perspective views "sickness as sin" applies in the most literal sense with regard to AIDS.

It is indicative of the extreme politicization of AIDS, especially in the United States, that so many of the questions that I have suggested as characteristic of health promotion policy debates remain unanswered. There is even disagreement over the seriousness of the problem itself, or at least on how to portray it. Some critics suggest that the degree of attention devoted to AIDS, and the implications of this for resource allocations, are disproportionate to the scope of the problem. Even those who are sympathetic to the plight of those afflicted with AIDS go to great lengths not to paint too bleak a picture of the disease for fear of creating rampant homophobia. Still others continue to compare AIDS to the plague and portray it as a health problem of unprecedented scope in this century.

The ambivalence that characterizes discussion of the scope of the problem is also evident in the issue of its origins. There is widespread belief in and support for the notion that AIDS is the result of engaging in a limited number of objectionable and immoral activities. In this sense AIDS is a classic self-inflicted life-style disease. Yet there is reluctance on the part of many policymakers, especially in Britain, to portray the disease in terms of social pathology because, in this view, it would serve no useful public health purpose to do so and might well be counterproductive.

Closely related to the uncertainties surrounding the questions of evidence and origins is the question of the nature and extent of the harm that AIDS poses for the nonhomosexual, non-drug-using population. There are obviously substantial direct costs for the care and treatment of AIDS that must be shared by the general population. But, in addition, there is the question of the non-material externalities of the disease. Does it pose a public health danger to non-high-risk groups? If not, then there might well be less of an inclination to commit substantial resources to the battle against AIDS, particularly given the competition over scarce funds among health professionals. On the other hand, if there is a real danger of a break out into the heterosexual community, then there will be greater pressure for extraordinary policy responses including universal testing, tracking, and so on.

And it is on the question of how to deal with the medical and legal implications of AIDS that these other questions come together. Most obviously in the United States, there is still no agreement on which of the various policy alternatives—education, tracking, testing, reporting, and so on—is the most appropriate. Here it is interesting to note that in general although conservatives favor the least intrusive response in other life-style issues, such as health education, in the case of AIDS they have championed the most aggressive use of government. For many conservatives, then, there is clearly a limit to how much foolishness the individual may engage in before government has a duty to intervene.

Finally, an observation on the impact of political structure on AIDS policy. One of Sandra Panem's recommendations for dealing with novel health emergencies like AIDS is to have responsibility and leadership "centralized and assigned to a federal official."[82] The British experience would seem to support this recommendation. However, such a response, while clearly superior, presupposes political and moral, as well as scientific, consensus on the origins, scope, and appropriate handling of the public health problem. Where such consensus does not exist, and it did not in the United States in the case of AIDS, decentralization may actually be beneficial, since it provides alternative opportunities for dealing with health problems. One shudders to think, for example, what might have happened if the British national leadership had been as split as the Reagan and Bush administrations were and are on the issue of how to respond to the tragedy posed by AIDS. Federalism in this instance has provided an avenue for education, innovation, and at least a fragmented response until some sort of national leadership consensus emerges.

[82] Panem, *The AIDS Bureaucracy*, 137.

Eight

Promoting Health and Protecting Freedom: American And British Experiences

No one wants to die of AIDS, lung cancer, cirrhosis of the liver, or injuries sustained in an automobile accident. The public policy debate is not over the desirability of avoiding illness, injury, or premature death, but over the individual and collective sacrifices we are willing to make to maximize our chances of living long and healthy lives. If those sacrifices were simply of a material nature, the personal and social dilemmas that life-style modification issues raise would be less intractable. The choice, however, is rarely limited to spending more or less money; it invariably involves allowing more or less personal freedom.

As I have emphasized throughout this book, health promotion policies touch upon some of the most essential and enduring issues of democratic theory and practice. Of particular relevance and importance are such questions as What is the appropriate role of the state in promoting responsible personal behavior? How, when they conflict, does the state reconcile the competing demands of personal freedom and collective well-being? There is implicit in these and related questions an image of two, incompatible models of state behavior. More than a century ago the British historian Thomas Babington Macaulay summarized these alternatives, and implicitly framed the dilemma, in the following way: "I hardly know which is the greater pest to society, a paternal government . . . which intrudes itself into every part of human life, and which thinks it can do everything better than anybody can do anything for himself, or a careless, lounging government, which suffers grievances, such as it could at once remove, to grow and multiply, and which to all complaint and remonstrance has only one answer: 'We must let things find their own level.' "[1]

Neither intrusive, albeit benevolent, state paternalism, nor benign neglect are acceptable models of state behavior in pluralist democracies like Britain and the United States. Instead, it is accepted, however grudgingly

[1] Thomas Babington Macaulay, *Speeches* (London: Longman, Brown, Green and Longman's, 1854), 436, 437, as quoted in Jack H. Schuster, "Out of the Frying Pan: The Politics of Education in a New Era," *Phi Delta Kappan* 63 (May 1982): 583.

by some, that the state has some obligation to regulate certain private choices in the public interest. Specifically with regard to health promotion, policymakers have struggled with practical and philosophical conditions and criteria to determine when and how government should act to ensure that individuals make more socially responsible life-style choices.

Because they share relatively comparable standards of living, levels of industrial and urban development, and morbidity and mortality profiles, both countries have many of the same public health problems. And, because they also share a common democratic ethos they each have had to reconcile demands for protecting both personal freedom and public health. However, while they start with similar political assumptions and social conditions, policymakers in each country have had to respond to major health challenges within a different cultural and institutional setting. The result has been that each has taken a somewhat different path toward dealing with some of the major health problems of the latter part of the twentieth century. In this concluding chapter I will do two things. First, I will assess some of the common behavioral standards that have emerged from debates over, and been applied to decisions concerning, life-style and health issues. Second, I will review the impact of the factors in my accounting scheme, and especially cultural predispositions and institutional arrangements, on health policy choices in Britain and the United States.

Where Do You Draw the Line?

Perhaps the most perplexing, and certainly the most frequently expressed, concern in the life-style and health debates examined in this book is over where one draws the line between public and private responsibility for protecting health. Lawmakers on the political left and right in both countries have relied on the standard of negative externalities as one guide to this elusive boundary. The typical conservative position has been that when a putatively injurious activity imposes physical harm, material costs, or significant inconvenience only on those engaged in the activity then it is outside the purview of the state. Those on the political left, on the other hand, have portrayed such choices as smoking cigarettes or driving without a seat belt as both harm- and cost-sharing behavior and, therefore, legitimate areas of public regulation.

To use negative externalities as a standard for judging if and when the state should exert jurisdiction over personal behavior is intuitively appealing. Surely if what I do harms others, then the state's obligation to protect the general welfare clearly supercedes my own right to be free to choose how I drive a motor vehicle, or when and where I smoke a cigarette or consume alcohol, or what sorts of sexual activities I engage in. Similarly, if I choose to be foolish, but that choice harms no one but myself, then a democratic state should be enjoined from interfering in my private affairs. The appeal and relevance of this standard has been illustrated repeatedly in health promotion policy debates in both countries. Certainly the question of externalities is at the heart of the tobacco industry's reaction to the issue of passive smoking. The assertion that environmental tobacco smoke poses a health hazard to the nonsmoker as well as the smoker is correctly seen by the tobacco interests as redefining the nature of the problem and, hence, the range of acceptable policy alternatives.

The problem with the concept of externalities as a guide to policy-making is that it is almost meaningless; virtually *every* form of behavior imposes *some* costs on those not directly involved in it. There are social costs, for example, even in an automobile accident involving a single motorist. This is especially true today, when so much of health care in both countries is publicly financed. The collective implications are further broadened if one includes indirect consequences of imprudent behavior, including those affecting social and private insurance, public safety costs, and economic productivity. In addition, some have expanded externalities through a type of displacement argument; that is, those who harm themselves through foolish behavior consume finite resources such as hospital beds or body organs for transplantation that could be used by those who become ill through no fault of their own.

Because it is potentially so broad, the utility of the externality standard as a guide to policymaking may depend ultimately upon specifying what types of costs or harms justify state intervention. One possibility is to make the distinction between physical and material externalities. Does a particular behavior pattern pose a physical danger to others, or does it merely impose material costs so that the prudent must subsidize the foolish? Or does it do both? Overindulgence in foods high in cholesterol may harm those who unwisely consume such foods and may impose costs on society through the use of health care facilities or increased insurance costs, but it does not physically endanger others. This distinction was critical in the debate over motorcycle helmet laws, in that defenders of the laws relied heavily on the shared costs, rather than the shared risks, of the failure to use a helmet. Although I cannot prove it, it seems to me that the belief that failure to wear helmets incurred only limited collective risks made it more difficult to sustain helmet laws in the United States, despite the collective costs. Any activity that poses a clear physical danger to others is, I would argue, a legitimate and appropriate state concern.

The next question, then, is whether activities that impose substantial material cost on others, without any attending physical harm, justify control through regulation. Here I would suggest that one possible answer is that it depends on whether these costs can be recaptured in some way so as to avoid the necessity of regulation and perhaps limiting freedom. In this regard, many have proposed levying extraordinary health taxes on cigarettes and alcohol. Revenue from these "sin taxes" would be designated to help pay the public bill to treat those with cirrhosis of the liver or lung cancer, in much the same way as motor-vehicle taxes are set aside for road repairs. These taxes might be objected to on the grounds that (1) some of those who engage in these risky behaviors do not do so voluntarily, but by genetic predisposition or addiction, and it is therefore unfair to penalize them for something over which they have no control; (2) the taxes are more appropriate for a national health care system (as in Britain) than one that is heavily privately financed (as in the United States); and (3) in order to be fair the taxes would have to apply to all potentially harmful practices, including eating foods high in salt, cholesterol, or sugar.[2] These objections are valid, but specifically designated tobacco and alcohol taxes seem a less objectionable approach to the problem of material externalities than regulation. Taxation is especially appealing since there is strong evidence relating increased cost to decreased consumption, provided the cost increase is high enough. If one accepts that decreased consumption of alcohol and tobacco is desirable, than a punitive health tax would accomplish the dual purpose of defraying costs and discouraging consumption.

Closely related to the question of the nature of the externalities produced by certain behavior is that of the extent and likelihood of these externalities. The reader will remember Alan Stone's assertion that "an externality whose probable impact is restricted to one or a small group does not warrant public action."[3] Unlike some aspects of economic activity, Stone's primary concern, life-style issues often involve substantial numbers of people. Nevertheless, the question of the extent of potential harm or cost is relevant and has been raised for each of the issues discussed in this book. The threat that AIDS poses to the non-intravenous-drug-using, heterosexual population has been a major issue and continues to influence the debates over policy options. In the United States, where AIDS is a more serious and substantial public health problem, a consen-

[2] See, for example, Marcia Kramer, "Self-Inflicted Disease: Who Should Pay for Care?" *Journal of Health Politics, Policy and Law* 4 (Summer 1979): 138–41; and Robert Veatch, "Who Should Pay for Smokers' Medical Care?" *Hastings Center Report* (November 1974): 8–9.

[3] Alan Stone, *Regulation and Its Alternatives* (Washington, D.C.: Congressional Quarterly Press, 1982): 98.

sus has yet to emerge over the nature of the threat. As a result, U.S. public policy remains uncertain and tentative. On the other hand, the case for a minimum alcohol-purchasing age was based largely on the widespread conviction that the self-inflicted and social harm resulting from youthful drinking habits was substantial.

Still another question in search of an answer is what standards of evidence we must require before accepting freedom-restricting, health-promoting policies. Here too I would suggest that we can do no better than accept intuitively compelling evidence of substantial harm. It will be recalled that one Conservative member of Parliament justified his support of a seat-belt law, but his reluctance in other health promotion areas, because the evidence on seat belts was "overwhelming." Although "overwhelming" is no more rigorous or objective than "substantial" and "significant," it is a reasonable and workable standard. I would suggest, for example, that there is little controversy over the notion that drunk drivers pose a significant social threat, or that the social costs of hang-gliding are relatively insignificant. There are, to be sure, important areas of disagreement: Does passive smoking pose a "significant" danger to a large number of people? Clearly the tobacco companies, who have implicitly accepted the standard of externalities, have sought to undermine the validity and credibility of the evidence linking environmental tobacco smoke with harm to nonsmokers. It seems to me that if and when the evidence on this link becomes as "overwhelming" as that concerning the direct harm to smokers, then laws regulating smoking in public places will become near-universal in both the United States and Britain, and properly so.

In sum, I would suggest that when an activity poses an actual, as opposed to potential, significant harm or material cost to a substantial number of others, and that harm cannot be reduced or avoided through voluntary means, education, or noncoercive disincentives such as higher taxes or insurance rates, then regulation is justified. The reader may well object to the imprecision of guidelines such as "significant threat" and "substantial number of others," but I would suggest that such terms are routinely used to describe a multiplicity of threats to public health, safety, or security, ranging from drug abuse to nuclear war. I recognize that such imprecise adjectives imbue the concept of externalities with seemingly inexhaustible elasticity that might erode personal freedom on many fronts. Furthermore, such standards seem extraordinarily inappropriate as guides to making public law. Yet I would suggest that the continued widespread, albeit diminishing, use of cigarettes, a quarter of a century after reports on its harmful effects were made public, suggests that in democracies, freedom, even to be foolish, is well protected.

Slopes and Wedges

Having said this, I hasten to add that legitimate concerns about the cumulative implications of health promotion regulation remain and need to be addressed. One such concern is epitomized by the metaphor of the slippery slope. Although each of the public health issues raised in this book has been dealt with individually, in reality they have all been considered in close temporal proximity. It has only been within the last decade or so that legislators, often the very same ones each time, have been asked to require the use of seat belts and motorcycle helmets, and raise the minimum drinking age, and limit the conditions of the sale and advertising of alcohol, and restrict the use of cigarettes, and support free syringes for intravenous drug users. Over and over again these legislators have asked, "Where will it all end?" The concern with the seeming accumulation of freedom-inhibiting public policies is reinforced by the incrementalist arguments often used to defend these measures. One argument in support of adult seat-belt laws was that they were a practical and logical extension of existing child restraint legislation or mandatory use of seat belts on airplanes. In addition, during the British debate over motorcycle helmets, one opponent warned that the logic behind that measure could easily be applied to requiring the use of seat belts by motorists—which is precisely what happened.

It is, of course, perfectly legitimate to argue that if one measure is successful in saving lives and reducing injury or illness, then it is incumbent on the state to extend it to analogous situations. If, for example, limiting cigarette advertisements reduces cigarette consumption—something that is open to dispute—then limiting advertising of alcohol products too makes sense. But, as some have pointed out, irresponsible use of automobiles also causes social harm, and should not this product therefore be subject to promotional restrictions? While arguments ad absurdum can be dismissed as mere rhetorical devices, they reflect a genuine concern that these policy actions are related, and that each subsequent assertion of state authority is more likely to be accepted than the previous one. One fruitful area of future public policy research might very well be to test the validity of the slippery slope argument.

Crying Wolf

Yet another issue affecting both popular and official reactions to proposed life-style modification policy is the tendency to portray each problem as the most serious health threat since the plague. Road accidents have been characterized as "the most intractable challenge of the second half of the 20th century"; cigarette smoking as second only to "nuclear annihilation"; AIDS as a "peril to our entire species"; and alcohol abuse as "the major public health issue of our time." Ironically, this hyperbole has hurt the cause of health promotion both because it has not been taken seriously *and* because it has been taken too seriously. These contradictory reactions are evident in the responses to AIDS by both pundits and policymakers. Thus there have been expressions of skepticism such as that of Charles Krauthammer, whom I quoted in Chapter 7, that perhaps AIDS does not "merit its privileged position at the head of every line of human misfortunes that make claims on our resources, attention and compassion."[4] Actually, the skeptical reaction to AIDS probably goes beyond this particular health problem and reflects the cumulative and numbing impact of the seemingly inflated statements concerning a variety of health problems over the years. Not every health problem can be "the most serious of our time," yet virtually all of them are presented as such. At the same time AIDS has engendered quite a different reaction. Legislators in both the United States and Britain have felt compelled to defuse some of the hysteria generated by attempts to convince the public and policymakers of the disease's seriousness. The problem here is not that people will belittle the danger but rather believe it too well and demand that extraordinary actions be taken against those who may pose a threat to the public's health. There is thus the risk that we will become either anesthetized or hysterical in the face of the apocalyptic claims on behalf of a multitude of health problems, many of which originate in our own carelessness. Although the extreme reactions of either an excessively intrusive and paternalist government or a "careless, lounging" one are not viable, there is the potential that government will either do too much, and thereby substantially reduce individual freedom, or too little, and thereby unwittingly contribute to the personal tragedies and social costs associated with avoidable deaths and illnesses.

Accounting for British and American Health Promotion Policy

The health risks and problems discussed in this book raise complex and troubling political issues. Concerns with demarcating public and private responsibility for following healthy life-styles, and the seemingly ever-narrowing choices available to citizens in pursuit of pleasure, pose serious political dilemmas on both sides of the Atlantic. The decision on if and how to respond to pressing public health problems is determined by a

[4] *Portland Oregonian*, 15 June 1987.

variety of idiosyncratic, cultural, and structural factors. I have tried to demonstrate that differences, particularly in cultural instincts and institutional arrangements, often lead British and American lawmakers to different conclusions about how to deal with similar health issues. Yet despite these differences, because there is such intense public and private sharing of information between these two English-speaking nations, there has been a good deal of policy sharing as well. The result, then, is that there are both similarities and differences in the nature and content of British and American health promotion policy. Let me summarize some of these.

To begin, the diffusion of ideas between the United States and Britain has helped set the agenda, framed the parameters of debate, and suggested the content of policy in a number of life-style and health issues. In the case of AIDS, for example, although the United States faced a "novel health emergency," the British had the American experience, albeit a fairly brief one, to guide it. And, as I have shown in Chapter 7, British policymakers and advisors have indeed drawn heavily upon the American experience and constantly use it as their frame of reference. In some instances their goal is emulation, in others avoidance.

Policy diffusion ran in the opposite direction in the case of seat-belt legislation. In this instance the British were the innovators, and the fact that they had already adopted a mandatory seat-belt law, and could point to initial success in both implementation and reduction of fatalities and serious injuries, was often noted in the media and legislative debates in this country.

There is certainly reciprocity in policy diffusion between the two countries, but it is clear that in health promotion the movement of ideas is more often than not from the United States to Britain rather than the reverse. Bernard Levin's lament, which I quoted in Chapter 1, that in life-style modification policy "the rule is: what America does today, Britain does tomorrow," has more than a little truth to it. Levin predicted that just as the British seem to be following the American anti-smoking campaign—remember that both the idea and content of cigarette health warnings were carried from the United States to Britain by a visiting M.P.—so too would they soon join in the current American assault on Demon Drink. Levin suggested that it was just a matter of time before British liquor bottles contained health warning labels. On the basis of the experiences in related policy areas, I suspect that he is correct.[5] I also suspect, however, that should the British ultimately adopt alcohol warning labels or place limits on alcohol advertising, they will do so through voluntary agreement rather than formal regulation. And herein lies one

[5] Bernard Levin, "Enter the Tea-Totalitarian," *Times* (London), 28 January 1988, 16b.

of the important differences between British and American health promotion policy.

Policy diffusion has had a certain homogenizing impact on health promotion policy in these two countries, but it has not been extensive or consequential enough to mask the significant differences that emerge from the peculiar cultural inclinations and structural factors that characterize British and American political, social, and economic life. There is, to begin, the impact of political structure on public policy. As noted in Chapter 1, one of the enduring debates in political science is over the impact of political structure on public policy. There is an intuitively compelling assumption, subscribed to by many of the Anglophiles in American academia and journalism, that unitary, parliamentary systems, such as the British one, are more efficient and effective processors of policy than federal systems with a separation of powers. One conclusion of this study is that political structure has had a mixed impact on the processing of health promotion policy. Decentralized decision-making facilitated the passage of anti-smoking legislation in the United States, where the congressional tobacco lobby was able to block it at the center, but not in individual states, while centralized decision making inhibited passage in Britain, where there were no alternative political arenas. By way of contrast, centralized decision-making produced earlier adoption of mandatory seat-belt legislation in Britain, where there was no active automobile industry lobby, than in the United States. Centralization of political power in Britain has sometimes facilitated action, and at other times inhibited it; decentralization in the United States has meant that at times the entire population has not been uniformly subject to health- and safety-protecting laws, like seat-belt legislation, while at other times it has only been because of decentralization that any of the population has been protected at all. Certainly for me one of the most important revelations resulting from my research was the largely counterintuitive conclusion that centralized decision-making, using nonregulatory approaches to problem-solving, was not necessarily more effective, efficient, or benevolent than the seemingly more chaotic, disjointed, and coercive policy routines characteristic of the American federal system.

It is clear that when a national consensus exists among the political elite, as it did in the case of seat belts and in response to the AIDS epidemic, the centralization of authority facilitates adoption of policies aimed at dealing with a critical public health problem. The absence of such a consensus in the United States has impeded the development of a rational and effective national AIDS policy. It has thus been left up to the individual states, where there are enormous variations in the magnitude of the problem, resources, public attitudes, and political commitment, to deal with a problem that in many important respects could best be dealt with on a national scope. Indeed, the need for orchestrated, federal leadership in the battle against AIDS was called for in the April 1990 report to President Bush from the bipartisan National Commission on AIDS: "There is no question that there have been creative and often heroic efforts at every level of government, but coordination of these efforts is the missing link to an effective national strategy."[6] By way of contrast when no national consensus exists, or powerful interest groups exercise a virtual veto over national legislation, highly centralized political systems such as the British are far less able to respond to public health problems than federal systems. In the United States, local and state governments have taken the lead on a number of issues relating to alcohol abuse and cigarette use in which the federal government could or would not become involved. From a purely public health perspective, the nation would be better off with a national seat-belt law. Such a law would undoubtedly save additional lives in those remaining states without mandatory legislation. Of course, the absence of such national legislative uniformity will be viewed with either dismay or relief depending upon one's personal ideological preference.

The impact of political structure on the ability of these nations to respond to health problems has been varied. Centralization of authority has both facilitated and inhibited efforts to deal with pressing health problems in Britain, while American federalism has at times provided opportunities for policy innovation, experimentation, and stimulation, and at other times resulted in deleterious fragmentation of effort. I would close on this point, however, by reiterating a comment made in the context of my discussion on the AIDS epidemic: one shudders to think what would have happened if in the face of this unique health emergency the British elite had not arrived at a consensus.

Although political structure has had a differential impact on the course and content of health policy in the United States and Britain, political style, culture, and ideology have had more uniform and predictable consequences. As a number of authors have noted, there is a discernible and characteristic British style of regulatory policymaking. This style, which contrasts dramatically with the American approach, is characterized as being flexible, informal, gradual, cooperative, and often shielded from public scrutiny.[7] There is, of course, something terribly appealing about such an approach. Rather than open, acrimonious, and divisive debates over cleaning up the environment, or protecting the consumer, or improving public health, the British allegedly can accomplish much of this with-

[6] *New York Times*, 25 April 1990.

[7] See David Vogel, *National Styles of Regulation: Environmental Policy in Great Britain and the United States* (Ithaca: Cornell University Press, 1986), 220–25, 269–80.

out becoming enmeshed, as American policymakers so often are, in a highly charged political conflict. I have suggested, however, that, contrary to David Vogel's conclusions about the superiority of the British regulatory style over the formal and coercive American approach, this style does not invariably or necessarily lead to less conflict or more effective policy. Vogel, perhaps reflecting the regulatory bashing now popular in the United States, found that the British style of voluntary and consensual agreements between government and industry not only minimized political conflict but was just as effective as the mandatory controls historically favored in the United States. I found, however, that this was not uniformly the case in the health promotion policy field. With the exception of AIDS, health promotion policy-making in Britain has not been as amicable as Vogel found environmental policy-making. The debate over seat-belt legislation lasted nearly a decade and was marked by considerable public conflict and acrimony. In addition, I would question Vogel's conclusions concerning the effectiveness of the British "corporatist" and consensual policy-making style. The British, for the cultural reasons identified in this book, have been slow in responding to the public health dangers posed by alcohol abuse and tobacco use.

It is particularly in the case of the public health threat posed by tobacco use that my own findings diverge most sharply from Vogel's. First, rather than minimize conflict, the voluntary agreements have actually prolonged and perpetuated disputes over smoking and health policy. Because voluntary agreements between industry and government are typically of a short and finite duration, every time they expire, the battle among health advocacy groups, the media, parliamentary backbenchers, the government, and the tobacco and advertising industries is revisited anew.

Second, since these voluntary agreements are not backed by the force of law and carry no penalties they have often been circumvented. I doubt that there is anyone who would seriously suggest that voluntary agreements on television tobacco advertising in Britain have been as successful, or the debates over them less acrimonious, than the legislative ban in the United States. The first voluntary agreement in Britain was reached in 1964, and a quarter of a century later the issue of promoting cigarettes on television has yet to be resolved. Actually, there is considerable evidence, including some from the government's own recently formed Committee for Monitoring Agreements in Tobacco Advertising and Sponsorship, that the advertisers and tobacco companies have persistently thwarted the intent of the agreements.

Why the difference between two such closely related policy areas as environmental and health promotion policy? Here I return to a point raised earlier in this book: health promotion policies involving individual life-styles inspire greater personal passions and produce more political conflict than other areas of public policy because they impinge more directly on personal freedom, choice, intimacy, and morality.

There are few in London or Washington, of any political persuasion, who would disagree with the proposition that when voluntary agreements work they are far preferable to state-mandated and enforced regulations of personal and corporate behavior. My point is that they do not always work. Furthermore, in the context of American public policy, they are unlikely to work. Voluntary agreements presuppose a history of trust, mutuality of interest, and confidence in and cooperation among government, industry, and relevant interest groups. These conditions have been characteristic of the British since the nineteenth century but not the American political culture. Americans tend to favor formal arrangements, enshrined in law, precisely because we are skeptical of both alleged government benevolence and corporate altruism.

This point introduces yet another crucial difference between the British and American policy milieus. Although it has been gradually eroded during the Thatcher years, the British culture still supports the view of government as a benevolent protector of the public good. This produces and sustains a general deference toward authority that not only facilitates voluntary agreements but also has implications for the implementation and success of policy choices. In this context I have compared the 95 percent seat-belt use in Britain since passage of mandatory legislation with that in those U.S. states with similar laws, where the average is 46 percent, to suggest that British deference to and confidence in government leads to acceptance of even initially unpopular laws. This contrasts with the situation in the United States where an adversarial relationship exists between government and the people on life-style modification issues. Similarly, I suspect that homosexuals in Britain feel comparatively less threatened by government policies than their American counterparts. Various events during Margaret Thatcher's decade in office, including inner-city racial conflict, violent labor disputes, hooliganism at sports events, and rioting over a new poll tax in 1990 suggest that the deferential quality of the British political culture may be fraying at the edges. These events startle the British, and others, however, precisely because they are so uncharacteristic. It may be that historians will view the Thatcher years as ushering in, or perhaps ratifying, fundamental changes in British attitudes toward government and the political and policy processes. For the present, and certainly during the period when the policies discussed in this book were under consideration, trust rather than antagonism between citizens and government prevailed in Britain.

The British and Americans also entertain different policy-relevant social values. The more tolerant attitude toward homosexuality in Britain, as reflected in both public opinion polls and legislative protection of homosexual relations, has facilitated a clearer, more supportive policy response to AIDS than has been possible in the United States. Similarly, the important social role of alcohol and the pub in British society has inhibited alcohol control policies there, while resurgent prohibitionism and "priggishness," to use Bernard Levin's term, have furthered the cause of such policies in the United States.

Finally, some comments on the role of ideology. There is an inclination in both countries for political observers to look for, and find, ideological consistency in the responses of policymakers to public problems. By and large we expect conservatives to privatize problems and solutions, while we expect liberals to collectivize them. In reality, however, although there are clear ideological predispositions, the picture is not quite as clear as one might expect. I have noted, for example, that a large number of British Conservative members of Parliament voted for the seat-belt bill, thus ensuring its adoption. In the United States liberal groups such as the American Civil Liberties Union and the National Organization for Women have opposed health-promoting policies such as bans on alcohol and tobacco advertisements or warnings to pregnant women on the dangers of alcohol on libertarian grounds, while conservatives have been the most vociferous proponents on collectivist grounds of vigorous state involvement in the attack on AIDS. Furthermore, although liberals in general have supported paternalistic policies like seat-belt laws and restrictions on smoking in public places, some have decried the victim-blaming nature of these measures. Such concerns, however, tend to emanate more from liberal academics than policymakers.

Of the two "dilemmas," the conservative one, especially under Reagan and Thatcher, has been the more obvious, consequential and, in many ways, interesting. Its origins and implications are different for each country. The dilemma for British Conservatives is that they have both a paternalistic and a libertarian tradition; in contemporary parlance there are "wets" (paternalists) and "drys" (libertarians). For most of the years following World War II, the Conservative party was controlled largely by its more traditional, paternalistic wing. Both the concept of noblesse oblige, and the realities of postwar British politics and public opinion, made the party an often ardent supporter of many welfare-state programs. With the ascendancy of Thatcher, a somewhat selective libertarian to be sure, the "dry" wing of the party has tended to dominate Conservative party politics and policies. In 1981, however, just two years after Thatcher's first selection as prime minister, there was enough "wet" Conservative support to help pass the seat-belt law. This faction of the party has been all but driven from the government and is considerably weakened among Conservative backbenchers. Hence, for example, despite some backbench opposition, the Thatcher government opted for boosting free enterprise, and enhancing freedom of choice, with the passage of liberalized opening hours for pubs.

In the United States the conservative dilemma is of a different sort. Here there is no paternalistic tradition, but there is a strong moralistic one. Some American conservatives like Senators Hatch and Thurmond, and former Senator Hawkins, who would normally resist state interference in private enterprise, have been the most vocal advocates of health warnings on cigarette and alcohol products, or limitations on advertising these products. Similarly, it has been conservatives who have called for aggressive state involvement in controlling AIDS. American conservatives, then, tend to support those state regulations that promote or protect, as they see it, fundamental moral values: sobriety, sexual abstinence outside of marriage and fidelity within, and "wholesome" living in general. The importance of these values to the conservative world view is underscored by the fact that conservatives are willing, on some occasions, to compromise another vital value, namely minimalist government. One could almost feel the discomfort of Senator Hatch (see Chapter 5) as he justified his support for health warnings on alcoholic beverages, despite his concern "with Government over-regulation and redtape." In the future, American conservatives will continue to face this dilemma.

As the twentieth century moves towards its conclusion, transnational forces, most notably rapid advances in the technology of communications and transportation, will inexorably lead to the more rapid movement of people, products, ideas, and, as the case of AIDS demonstrates, microorganisms across national borders. This, in turn, will facilitate the emergence of common national health policy agendas. Although national policy responses to common public health problems will continue to bear the imprint of the unique cultural and structural environments in which they occur, we can expect increasing convergence of policy options and choices. If, as the cases in this book suggest, the likelihood of state involvement in health promotion is a function of the quality of the scientific evidence linking particular behavior patterns to disease, then we can expect that involvement to accelerate as this evidence is shared more widely and quickly.

Bibliography

THIS bibliography is intended to identify some of the more important and accessible contemporary and historical works in the general field of health promotion, and in the specific areas of health promotion policy discussed in this book. I have not included in this bibliography many of the primary sources, including personal interviews and legislative debates, upon which I relied heavily in my research.

Historical Works

Amos, Sheldon. *A Concise Statement of the Argument against the Contagious Diseases Acts of 1864, 1866, and 1869.* Manchester: A. Ireland and Company, 1871.

Batcheler, Henry. *The Injustice, Inutility, and Immortality of the Contagious Diseases Acts.* London: Dyer Brothers, 1878.

Beck, Ann. "Issues in the Anti-Vaccination Movement in England." *Medical History* 4 (October 1984): 310–21.

Brandt, Allen M. *No Magic Bullet: A Social History of Venereal Disease in the United States since 1880.* New York: Oxford University Press, 1985.

Brend, William. *Health and the State.* London: Constable and Company, 1917.

Brown, Dorothy. *The Case for the Acceptance of the Sheppard-Towner Act.* Washington, D.C.: National League of Women Voters, 1923.

Brown, Ray A. "Police Power—Legislation for Health and Personal Safety." *Harvard Law Review* 42 (May 1929): 866–98.

Burnham, John C. "The Progressive Era Revolution in American Attitudes toward Sex." *The Journal of American History* 59 (March 1973): 885–908.

Charlton, Christopher. "The Fights against Vaccination: The Leicester Demonstration of 1885." *Local Population Studies* 30 (1983): 60–66.

Creighton, Charles. *A History of Epidemics in Britain.* 2d ed. New York: Barnes and Noble, 1965.

Duff, John. *A History of Public Health in New York City, 1866–1966.* New York: Russell Sage Foundation, 1964.

Duffy, John. "Social Impact of Disease in the Late Nineteenth Century." In *Sickness and Health in America*, edited by Judith Walzer Leavitt and Ronald L. Numbers. Madison: University of Wisconsin Press, 1978.

Fawcett, Millicent. "The Vaccination Act of 1898." *Contemporary Review* 75 (1899): 328–42.

Finer, S. E. *The Life and Times of Sir Edwin Chadwick.* London: Methuen and Company, 1952.

Frazer, W. M. *A History of English Public Health, 1834–1939.* London: Bailliere, Tindall and Cox, 1950.

Freymann, John Gordon. "Medicine's Great Schism: Prevention vs. Cure: An Historical Interpretation." *Medical Care* 13 (July 1975): 525–36.

Friedman, Lawrence J., and Arthur H. Shaffer. "History, Politics, and Health in Early American Thought: The Case of David Ramsay." *American Studies* 13 (1979): 37–56.

Gilbert, Bentley B. "Health and Politics: The British Physical Deterioration Report of 1904." *Bulletin of the History of Medicine* 39 (March–April 1965): 143–53.

Haley, Bruce. *The Healthy Body and Victorian Culture*. Cambridge: Harvard University Press, 1978.

Hodgkinson, Ruth G. "Provision for Health and Welfare in England in the Nineteenth Century." *Proceedings of the XXIII International Congress of the History of Medicine, London, 2–9* September 1972: 176–96.

Kaufman, Martin. "The American Anti-Vaccinationists and Their Arguments." *Bulletin of the History of Medicine* 41 (1967): 463–78.

Kramer, Howard D. "The Beginnings of the Public Health Movement in the United States." *Bulletin of the History of Medicine* 21 (March–April 1947): 352–76.

Lemons, J. Stanley. "The Sheppard-Towner Act: Progressivism in the 1920s." *Journal of American History* 55 (March 1969): 776–86.

Marcus, Alan I. "Disease Prevention in America: From a Local to a National Outlook, 1880–1910." *Bulletin of the History of Medicine* 53 (Summer 1979): 184–203.

Moore, Harry H. *Public Health in the United States*. New York: Harper and Brothers, 1923.

Newsholme, Sir Arthur. *The Last Thirty Years in Public Health: Recollections and Reflections on My Official and Post-Official Life*. London: George Allen and Unwin, 1936.

Paget, Major J. B. "The Health of the Nation." *English Review* 38 (March 1924): 377–85.

Ringen, Knut. "Edwin Chadwich, the Market Ideology, and Sanitary Reform: On the Nature of the Nineteenth-Century Public Health Movement." *International Journal of Health Services* 9 (1979): 107–20.

Rosen, George. "Disease, Debility, and Death." In *The Victorian City: Images and Reality*, edited by H. J. Dyos and Michael Wolf, vol. 2, 625–67. London: Routledge and Kegan Paul, 1973.

———. *A History of Public Health*. New York: MD Publications, 1958.

Rosenberg, Charles E. "The Bitter Fruit: Heredity, Disease, and Social Thought in Nineteenth-Century America." *Perspectives in American History* 8 (1974): 189–235.

———. *The Cholera Years: The United States in 1832, 1849 and 1896*. Chicago: University of Chicago Press, 1968.

Rosenkrantz, Barbara Gutmann. "Damaged Goods: Dilemmas of Responsibility for Risk." *Milbank Memorial Fund Quarterly* 57 (1979): 1–37.

———. *Public Health and the State: Changing Views in Massachusetts, 1842–1936*. Cambridge: Harvard University Press, 1972.

Ross, Dale L. "Leicester and the Anti-Vaccination Movement, 1853–1889." *Leicestershire Archaeological and Historical Society* 43 (1967–68): 35–44.

Schlesinger, Edward R. "The Sheppard-Towner Era: A Prototype Case Study in Federal-State Relationships." *The American Journal of Public Health* 57 (June 1967): 1034–40.

Shyrock, Richard H. "The Health of the American People: An Historical Survey." *Proceedings of the Philosophical Society* 90 (1929): 251–58.

———. "The Origins and Significance of the Public Health Movement in the United States." *Annals of Medical History* 1 (January 1929): 645–65.

Siefert, Kristine. "An Exemplar of Primary Prevention in Social Work: The Sheppard-Towner Act of 1921." *Social Work in Health Care* 9 (Fall 1983): 87–103.

Sigerist, Henry E. *Medicine and Human Welfare*. New Haven: Yale University Press, 1941.

Smith, F. B. "Ethics and Disease in the Later Nineteenth Century: The Contagious Disease Acts." *Historical Studies* 15 (1971): 118–35.

———. *The People's Health, 1830–1910*. New York: Holmes and Meier, 1979.

Stern, Bernard J. *Should We Be Vaccinated?* New York: Harper and Brothers, 1927.

Stewart, Alexander P., and Edward Jenkins. *The Medical and Legal Aspects of Sanitary Reform*. London: Robert Hardwick, 1867. Reprint. Leicester: Leicester University Press, 1969.

Stokes, John H. *Today's World Problem in Disease Prevention*. Ottawa: F. A. Acland, 1923.

Tobey, James A. "The Constitutionality of the Federal Maternity and Infancy Act." *Public Health Nurse* 14 (October 1922): 509–11.

———. *The National Government and Public Health*. Baltimore: Johns Hopkins University Press, 1926.

United Kingdom Fourth and Fifth Reports from the Royal Commission on Vaccination. Vol. 10. 1890–92.

United Kingdom Interdepartmental Committee on Physical Deterioration. *Report of the Inter-Departmental Committee on Physical Deterioration*. Cd. 2175. London: Her Majesty's Stationery Office, 1904.

United Kingdom Report from the Select Committee on the Contagious Disease Acts. Vol. 7. 1882.

Walkowitz, Judith R. *Prostitution and Victorian Society: Women, Class, and the State*. Cambridge: Cambridge University Press, 1980.

Watkin, Brian. *Documents on Health and Social Services: 1834 to the Present Day*. London: Methuen and Company, 1975.

Weeks, Jeffrey. *Sex, Politics and Society*. London: Longman, 1981.

White, Benjamin. *Smallpox and Vaccination*. Cambridge: Harvard University Press, 1925.

Williams, J. H. Harley. *A Century of Public Health in Britain, 1832–1929*. London: A. and C. Black, 1932.

Wohl, Anthony S. *Endangered Lives: Public Health in Victorian Britain*. Cambridge: Harvard University Press, 1983.

Contemporary Works

Allegrante, J. P., and R. P. Sloan. "Ethical Dilemmas in Workplace Health Promotion." *Preventive Medicine* 15 (1986): 313–20.

Bayer, Ronald, and Jonathan D. Moreno. "Health Promotion: Ethical and Social Dilemmas of Government Policy." *Health Affairs* 5 (Summer 1986): 72–85.

Beauchamp, Tom L. "The Regulation of Hazards and Hazardous Behaviors." *Health Education Monographs* 6 (Summer 1978): 242–57.

Berlinger, Giovanni. "Life-Styles and Health: Alternative Patterns." *International Journal of Health Services* 11 (1981): 53–61.

Brady, John Paul. "Behavior, the Environment, and the Health of the Individual." *Preventive Health* 12 (September 1983): 600–609.

Califano, Joseph A., Jr. *America's Health Care Revolution.* New York: Random House, 1986.

Callahan, Daniel. "Preventing Disease, Creating Society." *American Journal of Preventive Medicine* 2 (1986): 205–8.

Carr-Hill, Roy. "The Inequalities in Health Debate: A Critical Review of the Issues." *Journal of Social Policy* 16 (no. 4) (1987): 509–42.

Cox, B. D., et al. *The Health and Lifestyle Survey.* London: Health Promotion Research Trust, 1987.

Crawford, Robert. "Individual Responsibility and Health Politics in the 1970s." In *Health Care in America*, edited by Susan Reverby and David Rosner. Philadelphia: Temple University Press, 1979.

———. "You Are Dangerous to Your Health: The Ideology and Politics of Victim Blaming." *International Journal of Health Services* 7 (1977): 663–80.

Epstein, Samuel S., and Joel B. Swartz. "Fallacies of Lifestyle Cancer Theories." *Nature* 289 (January-February 1981): 127–30.

Evans, Robert. "A Retrospective on the 'New Perspective.' " *Journal of Health, Politics, Policy and Law* 7, no. 2 (Summer 1982): 325–44.

Fuchs, Victor. "The Economics of Health in a Post-Industrial Society." *The Public Interest*, no. 56 (Summer 1979): 3–20.

Haggarty, Robert J. "Changing Lifestyles to Improve Health." *Preventive Medicine* 6 (1977): 276–89.

Halsey, A. H. *Change in British Society.* Oxford: Oxford University Press, 1986.

Harris, Daniel M., and Sharen Guten. "Health-Protective Behavior: An Exploratory Study." *Journal of Health and Exploratory Behavior* 20, no. 1 (1979): 17–29.

Holtzman, Neil A. "Prevention: Rhetoric and Reality." *International Journal of Health Services* 9 (1979): 25–39.

Imperato, Pascal James, and Greg Mitchell. *Acceptable Risks.* New York: Viking, 1985.

Kass, Leon R. "Regarding the End of Medicine and the Pursuit of Health." *Public Interest* no. 40 (Summer 1975): 11–42.

Kennedy, Donald. "Health, Science, and Regulation: The Politics of Prevention." *Health Affairs* 2 (Fall 1983): 39–51.

Knowles, John H., ed. *Doing Better and Feeling Worse: Health in the United States*. New York: Norton, 1977.

Lalonde, Marc. *A New Perspective on the Health of Canadians*. Ottawa: Department of National Health and Welfare, 1974.

McDermott, Walsh. "Medicine: The Public Good and One's Own." *Perspectives in Biology and Medicine* 21, no. 2 (Winter 1978): 167–87.

McGinnis, J. Michael. "Analyzing the Moral Issues in Health Promotion." *Preventive Medicine* 10 (1981): 379–81.

———. "The Limits of Prevention." *Public Health Reports* 100 (May–June 1985): 255–60.

McKeown, Thomas. *The Role of Medicine: Dream, Mirage, or Nemesis*. Princeton: Princeton University Press, 1979.

Mencher, Samuel. "Trends in Government's Health Roles." *Bulletin of the New York Academy of Medicine* 41 (December 1965): 1350–58.

Neubauer, Deane, and Richard Pratt. "The Second Public Health Revolution: A Critical Appraisal." *Journal of Health Politics, Policy and Law* 6 (1981): 205–28.

The 1990 Health Objectives for the Nation: A Midcourse Review. Washington, D.C.: Public Health Service, 1986.

Pellegrino, Edmund D. "Health Promotion as Public Policy: The Need for Moral Groundings." *Preventive Medicine* 10 (May 1981): 371–78.

Perkins, Richard J. "Perspectives on the Public Good." *American Journal of Public Health* 71 (March 1981): 294–97.

Pollard, Michael R., and John T. Brennan. "Disease Prevention and Health Promotion Initiatives: Some Legal Considerations." *Health Education Monographs* 6 (1978): 211–22.

Robbins, Christopher. *Health Promotion in North America: Implications for the U.K.* London: Health Education Council and King Edward's Hospital Fund, 1987.

Rogers, Peggy Jean, Elizabeth K. Keaton, and John G. Bruhn. "Is Health Promotion Cost Effective?" *Preventive Medicine* 10 (1981): 324–39.

Russell, Louise. *Is Prevention Better than Cure?* Washington, D.C.: Brookings Institution, 1986.

Saward, Ernest, and Andrew Sorenson. "The Current Emphasis on Preventive Medicine." In *Health Care: Regulation, Economics, Ethics, Practice*, edited by Philip H. Abelson, 49–54. Washington, D.C.: Association for the Advancement of Science, 1978.

Somers, Anne R. "Rights and Responsibilities in Prevention." *Health Education Monographs* 6 (1978): 37–39.

Stone, Deborah A. "At Risk in the Welfare State." *Social Research* 56, no. 3 (Autumn 1989): 591–633.

———. "The Resistible Rise of Preventive Medicine." *Journal of Health Politics, Policy and Law* 11 (1986): 671–96.

Taylor, Rosemary C. R. "The Politics of Prevention." *Social Policy* 13 (1982): 32–41.

Taylor, Rosemary C. R. "State Intervention in Postwar European Health Care: The Case of Prevention in Britain and Italy." In *The State in Capitalist Europe*, edited by Stephen Bornstein, David Held, and Joel Krieger, 91–111. London: George Allen and Unwin, 1984.

Tesh, Sylvia. "Disease Causality and Politics." *Journal of Health Politics, Policy and Law* 6, no. 3 (Fall 1981): 369–90.

United Kingdom Department of Health and Social Security. *Prevention and Health*. London: Her Majesty's Stationery Office, 1977.

———. *Prevention and Health: Everybody's Business*. London: Her Majesty's Stationery Office, 1976.

United States Department of Health, Education, and Welfare. *Healthy People: The Surgeon General's Report on Health Promotion and Disease Prevention*. Washington, D.C.: Government Printing Office, 1979.

United States Department of Health and Human Services. *Promoting Health/Preventing Disease: Objectives for the Nation*. Washington, D.C.: Government Printing Office, 1980.

United States Surgeon General. *Healthy People*. Washington, D.C.: Public Health Service, 1979.

Veatch, Robert M. "Voluntary Risks to Health: The Ethical Issues." *Journal of the American Medical Association* 243 (January 4, 1980): 50–55.

Weale, Albert. "Invisible Hand or Fatherly Hand? Problems of Paternalism in the New Perspetive on Health." *Journal of Health Politics, Policy and Law* 7 (Winter 1983): 784–807.

Wikler, Daniel I. "Persuasion and Coercion for Health: Ethical Issues in Government Efforts to Change Life-Styles." *Milbank Memorial Fund Quarterly* 56, no. 3 (1978): 303–38.

Wilkinson, J. H., III, and G. E. White. "Constitutional Protection for Personal Lifestyles." *Cornell Law Review* 62 (March 1977): 563–620.

Yankauer, Alfred. "Prevention and the Public Health." *Preventive Medicine* 13 (1984): 323–26.

Smoking and Health

Brenslow, Lester. "Control of Cigarette Smoking from a Public Policy Perspective." *Annual Review of Public Health* 3 (1982): 129–51.

Calnan, Michael. "The Politics of Health: The Case of Smoking Control." *Journal of Social Policy* 13 (July 1984): 279–95.

Corina, Maurice. *Trust in Tobacco: The Anglo-American Struggle for Power*. London: Michael Joseph, 1975.

Diehl, Harold S. *Tobacco and Your Health: The Smoking Controversy*. New York: McGraw-Hill, 1969.

Drew, Elizabeth. "The Quiet Victory of the Smoking Lobby." *Atlantic Monthly* 216 (September 1965): 76–80.

Finger, William R., ed. *The Tobacco Industry in Transition*. Lexington, Massachusetts: D.C. Heath and Company, 1981.

Fiore, Michael C., Thomas E. Novotny, John D. Pierce, Evridiki J. Hatziandreu, Kantilal M. Patel, and Ronald M. Davis. "Trends in Cigarette Smoking in the

United States: The Changing Influence of Gender and Race." *Journal of the American Medical Association* 261 (January 6, 1989): 49–55.

Friedman, Kenneth M. *Public Policy and the Smoking-Health Controversy*. Toronto: Lexington Books, 1975.

Fritschler, A. Lee. *Smoking and Politics*. 3d ed. Englewood Cliffs, New Jersey: Prentice-Hall, 1983.

Jacobsen, Bobbie. *The Ladykillers: Why Smoking Is a Feminist Issue*. London: Pluto Press, 1981.

Neuberger, Maurine B. *Smoke Screen: Tobacco and the Public Welfare*. Englewood Cliffs, New Jersey: Prentice-Hall, 1963.

Ravenholt, R. T. "Tobacco's Impact on Twentieth-Century Morality Patterns." *American Journal of Preventive Medicine* 1, no. 4 (1985): 4–17.

Royal College of Physicians. *Health or Smoking?* London: Pitman Publishing, 1983.

———. *Smoking and Health*. London: Pitman Publishing, 1962.

———. *Smoking or Health?* London: Pitman Publishing, 1977.

Sapolsky, Harvey M. "The Political Obstacles to the Control of Cigarette Smoking in the United States." *Journal of Health Politics, Policy and Law* 5 (Summer 1980): 277–90.

Taylor, Peter. *Smoke Ring: The Politics of Tobacco*. London: The Bodley Head, 1984.

United States Surgeon General. *The Health Consequences of Involuntary Smoking*. Washington, D.C.: Government Printing Office, 1986.

———. *The Health Consequences of Smoking: Cancer*. Washington, D.C.: Department of Health and Human Services, 1982.

———. *The Health Consequences of Smoking for Women*. Washington, D.C.: Department of Health and Human Services, 1980.

———. *Reducing the Health Consequences of Smoking: Twenty-five Years of Progess*. Washington, D.C.: Department of Health and Human Services, 1989.

Wagner, Susan. *Cigarette Country: Tobacco in American History and Politics*. New York: Praeger Publishers, 1971.

Warner, Kenneth E. "Cigarette Advertising and Media Coverage of Smoking and Health." *New England Journal of Medicine* 312 (February 7, 1985): 384–88.

———. "Smoking and Health Implications of a Change in the Federal Cigarette Excise Tax." *Journal of the American Medical Association* 255 (February 28, 1986): 1028–32.

———. "State Legislation on Smoking and Health: A Comparison of Two Policies." *Policy Science* 13 (April 1981): 139–52.

White, Larry C. *Merchants of Death: The American Tobacco Industry*. New York: Beech Tree Books, 1988.

Alcohol Abuse

Cavanagh, John, and Frederick F. Clairmonte. *Alcoholic Beverages: Dimensions of Corporate Power*. New York: St. Martin's, 1985.

Colvin, Michael. *Time Gentlemen Please*. London: Conservative Political Centre, 1986.

De Lint, Jan E. E. "Alcohol Control Policy as a Strategy of Prevention." *Journal of Public Health Policy* 1 (March 1980): 41–49.

Harrison, Brian. *Drink and the Victorians: The Temperance Question in England, 1815–1872.* Pittsburgh: University of Pittsburgh Press, 1971.

Kendell, R. E. "Alcoholism: A Medical or Political Problem? *British Medical Journal* 1 (February 10, 1979): 367–71.

Lender, Mark E., and James K. Martin. *Drinking in America: A History.* New York: The Free Press, 1982.

Lewis, John. *Freedom to Drink.* London: Institute of Economic Affairs, 1985.

National Alcohol Tax Coalition. *Impact of Alcohol Excise Tax Increases on Federal Revenues, Alcohol Consumption, and Alcohol Problems.* Washington, D.C.: National Alcohol Tax Coalition, 1985.

National Institute on Alcohol Abuse and Alcoholism. *Alcohol and Health.* Washington, D.C.: United States Department of Health and Human Services, 1981.

Paton, Alex. "The Politics of Alcohol." *British Medical Journal* 290 (January 5, 1985): 1-2.

Royal College of Psychiatrists. *Alcohol and Alcoholism.* London: Tavistock Publications, 1979.

———. *Alcohol: Our Favorite Drug.* London: Tavistock Publications, 1986.

Spiegel, David R. "Regulating Demon Rum." *Regulation: AEI Journal on Government and Society* 9 (March-April 1985): 43–48.

United Kingdom Advisory Committee on Alcoholism. *Report on Prevention.* London: Her Majesty's Stationery Office, 1977.

United Kingdom Department of Health and Social Security. *Prevention and Health: Drinking Sensibly.* London: Her Majesty's Stationery Office, 1981.

United States Congress. Senate. Subcommittee on Children, Family, Drugs, and Alcoholism. *Alcohol Advertising.* Hearings, 1985.

United States Presidential Commission on Drunk Driving. *The New Hope of Solution.* Washington, D.C.: 1983.

Wilson, George B. *Alcohol and the Nation.* London: Nicholson and Watson, 1940.

Automobile Safety

Claybrook, Joan B. "Changing Life-Styles, Health, and Highway Safety." *Preventive Medicine* 8 (1979): 112–15.

Crandall, Robert W., Howard K. Gruenspecht, Theodore E. Keeler, and Lester B. Lave. *Regulating the Automobile.* Washington, D.C.: Brookings Institution, 1986.

Graham, John D., and Patricia Gorham. "NHTSA and Passive Restraints: A Case of Arbitrary and Capricious Deregulation." *Administrative Law Review* 35 (Spring 1983): 193–252.

Leichter, Howard M. "Lives, Liberty and Seat Belts in Britain: Lessons for the United States." *International Journal of Health Services* 16., no. 2 (1986): 213–25.

———. "Saving Lives and Protecting Liberty: A Comparative Study of the Seat-

Belt Debate." *Journal of Health Politics, Policy and Law* 11, no. 2 (Summer 1986): 323–44.

MacKay, Murray. "Seat Belts and Risk Compensation." *British Medical Journal* 291 (September 21, 1985): 757–58.

Nader, Ralph. *Unsafe at Any Speed: The Designed-In Dangers of the American Automobile*. New York: Grossman Publishers, 1965.

Peltzman, Sam. *Regulation of Automobile Safety*. Washington, D.C.: American Enterprise Institute, 1975.

United Kingdom Department of Transport. *Interdepartmental Review of Road Policy Safety*. London: Her Majesty's Stationery Office, 1987.

———. *Road Safety: The Next Steps*. London: Her Majesty's Stationery Office, 1987.

United Kingdom Parliament Transport Committee. "Second Special Report from the Transport Committee: Road Safety." Commons, 1983.

Warner, Kenneth E. "Bags, Buckles, and Belts: The Debate over Mandatory Passive Restraints in Automobiles." *Journal of Health Politics, Policy and Law* 8 (Spring 1983): 44–75.

AIDS

Altman, Dennis. *AIDS in the Mind of America*. Garden City, New York: Anchor Press, 1986.

———. "The Politics of AIDS." In *AIDS: Public Policy Dimensions*, edited by John Griggs. New York: United Hospital Fund, 1987.

Bateson, Mary Catherine, and Richard Goldsby. *Thinking AIDS*. Reading, Massachusetts: Addison-Wesley, 1988.

Bayer, Ronald. "AIDS and the Gay Community: Between the Specter and the Promise of Medicine." *Social Research* 52 (Autumn 1985): 581–606.

———. "AIDS, Power and Reason." *The Milbank Quarterly* 64, supplement 1 (1986): 168–82.

Black, David. *The Plague Years*. New York: Simon and Schuster, 1986.

Brandt, Allan M. "The Syphilis Epidemic and Its Relation to AIDS." *Science* 239 (January 22, 1988): 375–80.

Cahill, Kevin M., ed. *The AIDS Epidemic*. New York: St. Martin's, 1983.

Fleming, Alan F., et al., eds. *The Global Impact of AIDS*. New York: Alan R. Iss, 1988.

Fletcher, James. "Homosexuality: Kick and Kickback." *Southern Medical Journal* 77 (February 1984): 149–50.

Fox, Daniel M., Patricia Day, and Rudolf Klein. "The Power of Professionalism: Policies for AIDS in Britain, Sweden, and the United States." *Daedalus* 118, no. 2 (Spring 1989): 93–112.

Gostin, Larry, and William J. Curran. "The Limits of Compulsion in Controlling AIDS." *Hastings Center Report* 16 (December 1986): 24–29.

House of Commons Social Services Committee. *Problems Associated with Aids*. Session 1986–87. London: Her Majesty's Stationery Office, 1987.

Hyde, H. Montgomery. *The Love That Dare Not Speak Its Name*. Boston: Little, Brown and Company, 1970.

Kirby, Michael. "AIDS and Law." *Daedalus*, 118, no. 3 (Summer 1989): 101–21.

Kubler-Ross, Elisabeth. *AIDS: The Ultimate Challenge*. New York: Macmillan, 1987.

Langone, John. *AIDS: The Facts*. Boston: Little, Brown and Company, 1988.

Macklin, Ruth. "Predicting Dangerousness and the Public Health Response to AIDS." *Hastings Center Report* 16 (December 1986): 16–23.

Merritt, Deborah Jones. "The Constitutional Balance between Health and Liberty." *Hastings Center Report* 16 (December 1986): 1–10.

Mills, S., M. J. Campbell, and W. E. Waters. "Public Knowledge of AIDS and the DHSS Advertisement Campaign." *British Medical Journal* 293 (October 25, 1986): 1089–90.

Panem, Sandra. *The AIDS Bureaucracy*. Cambridge: Harvard University Press, 1988.

Shilts, Randy. *And the Band Played On*. New York: St. Martin's Press, 1987.

Silverman, Mervyn, and Deborah Silverman. "AIDS and the Threat to Public Health." *Hastings Center Report* 15 (August 1985): 19–22.

Sontag, Susan. *AIDS and Its Metaphors*. New York: Farrar, Straus and Giroux, 1988.

United Kingdom Department of Health and Social Security. *AIDS and Drug Misuse Part I: Report by the Advisory Council on the Misuse of Drugs*. London: Her Majesty's Stationery Office, 1988.

———. *Future Trends in AIDS*. London: Her Majesty's Stationery Office, 1987.

———. *Problems Associated with AIDS: Response by the Government to the Third Report from the Social Services Committee*. Session 1986–87. London: Her Majesty's Stationery Office, 1988.

The Wolfenden Report: Report of the Committee on Homosexual Offenses and Prostitution. Authorized American ed. New York: Stein and Day, 1963.

Index